Nature Is Screaming! Are We Ready to Listen?

Steve Conboy

Published by Creative Classics Publications US

Nature Is Screaming;
Are We Ready to Listen?

Copyright 2022 Steve Conboy

Printed in the United States of America

Book layout: Abraham Obafemi Emmanuel
Editor: Kathleen Tracy

ISBN: 978-1-7352724-7-4

▬

This book is in honor of Thomas Conboy, my late father who is in heaven. His life was cut short after being told in the Navy that the asbestos used to wrap pipes was safe. When it comes to protecting human and animal life, our waterways, and the future of our environment, we have to embrace change and adopt safer, more environmentally friendly chemistry to fight fires.

▬

CONTENTS

CONTENTS

FOREWORDS

When I first became aware of Steve Conboy and his fire inhibiting products, I was a serious doubting Thomas, who advised him I'd believe it when I saw it. After thirty-five years in the fire service, I thought I had seen it all, but after several demos of both products I knew that this was something new and a game-changer. I look forward to helping him convince others that it truly is for real. After reading this book I hope you too will be willing to see how it can change fire defense practices. As has been shown in the past several years, continuing to do the same things we have been doing won't get the job done. The future is now with Green Tech Wildfire Defense.

Jeff Bowman,
Retired California fire chief

I have watched Steve Conboy push forest product sustainability and create the most cost-effective fire protection for lumber the building industry has ever known. This guy has been listening to nature's screams for more than twenty years. As a structural engineer I fully support the way this technology is defending wood structures from fire because our industry has never made this type of protection so affordable. It makes the decision to use it easy.

Tom Curry
President, Performance Plus Engineering in Temecula, CA

PREFACE

My father, Thomas Conboy, was a Navy engineer and served at a time when it was deemed safe to wrap pipes with wet asbestos on US military ships. Back then we didn't know better. My father died forty years later from cancer. That loss has fueled my passion to protect workers, fire fighters, and others through safety technology, such as M-Fire's wildfire defense systems. This book is also intended to prompt a bigger discussion about the construction industry as a whole with an imperative to make it safer, greener, and more sustainable.

I believe we can have a real, positive impact on reversing the damage to our planet by engaging in four sustainable best practices: providing fire protection to defend the carbon stored in wood buildings, promoting additional reforestation, investing in clean solar and wind energy, and developing better and faster ways to proactively defend all our structures and wildlands from fires to shut down greenhouse gas production from those sources.

I feel it's my job and my purpose to reduce all fire risk so my grandchildren can live on a planet tomorrow that was as good as we once had it.

I hope the information in this book will help educate the public, government officials, public safety administrators, insurance companies, and construction professionals about new technology that can be implemented now. This applied fire science will make an immediate, positive impact on advancing affordable housing, environmental responsibility, and renewable sustainability.

THE MIGHTY FIRE BREAKER MISSION

When it comes to wildfires and homes we still follow Ben Franklin's sage advice that an ounce of prevention is worth a pound of cure. There is a growing body of best practices that have been proven to protect homes and businesses but none as clean and safe as Mighty Fire Breakers' proven accredited fire inhibitor. If the government would step in now with the appropriate policies and resources we could reduce the losses going forward significantly. If they were supported by a tax credit, we would get even more people to embrace real wildfire engineering that is not just counting on vegetation fuel trimming but more on better applied science.

We all know that wildfires have always been a natural feature of the US landscape but we have to take blame for building within those regions without any proven methods to defend our homes. Nature does a better job on its own with bark than we have done on homes.

All the catastrophic wildfires that have devastated cities and towns such as Paradise, Santa Rosa, and Redding in California must be the catalyst for driving us toward new and better applied science beyond just adding additional planes to dump water or fire retardant—a strategy that is not particularly effective and can have negative environmental consequences.

We also need to focus more on entire community defense systems and not just the forests, which have received most of the attention. It's time we start protecting communities in wildlands regions with new and better wildfire defense engineering. The government has to encourage innovation from the private sector to develop new methods of wildfire fighting that better defend the future of our environment than current policies and procedures do. Our environment has been harmed by toxic chemicals and the neglect of not adopting a truly sustainable approach supported by more technologically advanced applied fire science and practices.

Our mission at Mighty Fire Breaker with our patented Green Tech Wildfire Defense is to revolutionize wildfire defense as well as defending wood-framed buildings and help minimize environmental damage from oil and gas disasters.

Our company follows true sustainability practices because not only is it environmentally friendly, it's also good business and our products will become increasingly good investments for others. Every application in our portfolio has a positive impact on reducing greenhouse gas and CO_2, and I believe it will soon be worthy of a carbon trade, which is the buying and selling of credits that permit a company or other entity to emit a certain amount of carbon dioxide. The value of the carbon is based on the ability of the country to store it or to prevent it from being released into the atmosphere.

THE INTERTWINED HISTORY OF AMERICAN HOUSING BOOMS AND WILDFIRES

——

B efore the boom came the bust.

In 1932 at the height of the Great Depression, a quarter of a million people lost their property to foreclosure. By the end of 1933, half of all American mortgages were in default, with a foreclosure of than one thousand per day. The dire state of housing and homeownership prompted President Franklin Roosevelt to create the Home Owners' Loan Corporation (HOLC), which began buying and refinancing existing mortgages at risk of default.

HOLC introduced the amortized mortgage, allowing borrowers to pay back interest and principal over twenty to thirty years. Prior to that it was standard to offer a five-year mortgage with a balloon payment as the final payment. Though homeowners ultimately paid more for their homes with an amortized mortgage, their monthly payments were lower and consistent. The new system made it possible for many more people to become homeowners, which in turn provided increased personal stability and a pathway to build equity and accrue wealth as their property values rose over time.

The modern housing industry began in earnest following the end of World War II, when more than four million men and women serving in the armed forces came back home. With most looking to resume their civilian lives, the marriage rate soared, resulting in the biggest baby boom in United States history up to that time. The demand for housing quickly exceeded the supply, creating an opportunity for enterprising contractors, who found ample space in the outskirts of cities.

The first large, planned community, Levittown, was developed in New York by William Levitt, who's been called the father of American suburbia. Levitt developed an assembly line method of production using his experience in the Navy, where Levitt worked in constructing military housing. He used those methods to streamline construction of residential homes that featured then state-of-the-art technologies. The downside was they also tended to look like they came from the same cookie cutter, making them mostly devoid of personality, but his system was highly efficient and helped Levitt build more than seventeen thousand houses. Levitt might have been the first, but many followed and communities like Levittown sprang up across the country.

The GI Bill and Federal Housing Authority (FHA) loan programs subsidized low-cost mortgages, which meant that it was often cheaper to buy one of these suburban houses than it was to rent an apartment in the city. So American families migrated to suburbia, cementing our car culture, which in the following decades would lead to home construction in ever more remote locales, including in and around areas that have been burning since dinosaurs roomed this earth.

In the 1950s and 1960s, advances in air conditioning enabled a housing rush throughout the Sun Belt and in southern cities such Atlanta and Miami. Prior to World War II, Americans had predominantly lived in metropolitan areas. But in the post-war years, population growth occurred in the suburban areas, so cities, which had historically been the heart of society and culture, lost jobs and population while suburbia population swelled, going from 19.5 percent in 1940 to almost 31 percent by 1960. By the late 1960s, almost as many Americans lived in suburban areas as in city centers. Between 1950 and 1970, the United States' suburban population nearly doubled to seventy-four million.

Arguably the biggest housing boom began in early 1997 and lasted nine years. According to the *New York Times*, "This was a period of intense speculative enthusiasm—for houses and for financial instruments based on mortgages as investments—and it was also a time of great regulatory complacency. The term *flipping houses* became popular then. People exploited the boom by buying homes and selling them only months later at a huge profit."

That boom imploded in 2006, cut off at the knees by the subprime mortgage scandal, caused by banks selling too many high-risk mortgages to feed the demand for mortgage-backed securities after deregulation allowed banks to essentially act as hedge funds. Falling home prices in 2006 left many homeowners underwater—meaning they owed more on their mortgage than what the house was worth in the market—and triggered a tsunami of defaults, which led to the Great Recession of 2008–2010.

But housing has always proved resilient, and in early 2012 the industry bounced back. Construction of new houses increased, and banks started offering mortgage loans, albeit with much stronger lending requirements. The flourishing economy lasted through the start of 2020, until the coronavirus pandemic put the squeeze on many industries beginning that March. But housing remained robust, COVID be damned. In the first six months of 2020, Quicken reported a record $120 billion in home loans funded, the company's best lending year ever—and with six months left to go. New home starts—a term referring to the number of new residential construction projects that begin during any particular month—surged 17 percent in June 2020. their highest level since 2006. The number of building permits granted rose to levels not seen since March 2007.

As the pandemic wore on, in-person real estate showings dwindled amid public health officials calling for social distancing and municipalities issuing shelter-in-place directives. But real estate agents quickly pivoted to using virtual tours to help clients buy and sell properties.

In January 2021, the *Wall Street Journal* reported, "Sales of previously owned homes rose in 2020 to the highest level since 2006, as ultralow interest rates and remote work during the pandemic increased homebuying demand."

In early 2021 demand had started to outpace supply, creating a situation reminiscent of post-World War II America. There are two reasons for this undersupply. First, Americans are holding on to their homes longer, and it is costing would-be home buyers. A report by real estate brokerage Redfin Corp found one in four American homeowners have lived in the same home for more than twenty.

The *Wall Street Journal* reported, "The length of time US homeowners stay put has been rising steadily, a big reason why the inventory of homes for sale is at record lows and prices are near all-time highs. The typical homeowner in 2020 had remained in place for thirteen years … well ahead of 2010's reading of 8.7 years."

The second reason for this undersupply is because millennials have started buying homes. According to the National Association of Realtors, four in ten homebuyers in 2019 was a millennial. And every year for the next decade a new generation of homebuyers will reach home buying age, tens of millions of millennials, requiring ever more new homes being built. While this is good for the housing industry, there is also a growing downside: where there are homes, there is an increased risk of people being endangered by a variety of natural disasters, especially catastrophic wildfires. The statistics are staggering:

From 2005 to 2020, wildfires have destroyed 89,210 structures, including homes and businesses. The wildfire seasons from 2017 to 2020 account for 62 percent of the structures lost over that fifteen-year span.

The 2018 Camp Fire was the deadliest and most destructive fire in California's recorded history. Eighty-five people died, more than 150,000 acres burned, and 18,800 structures were destroyed, resulting in losses upward of $10 *billion*, making it also the costliest wildfire on record.

A 2019 wildfire risk analysis identified 4.5 million US homes as being at high or extreme risk of wildfire. In 2021, more than two million properties in California were considered at high to extreme wildfire risk.

From January 1 to November 26, 2021, there were 52,729 wildfires, burning 6.6 million acres.

Beyond the physical damage such an inferno causes, it also releases dangerous gases into our atmosphere that exacerbate climate change, which in turn make additional wildfires that much more likely. This isn't a hypothetical; wildfires are a growing threat to both man and environment.

Take ozone. To measure the impact of the fires on ozone formation, researchers used computer models developed at the National Center for Atmospheric Research. With the first one, a specialized fire model, they estimated the amount of vegetation burned and the resulting emissions of nitrous oxides, sulfur dioxide, and other pollutants. Those results went into a global air chemistry model that simulated the movement of the emissions and evolving chemistry and tracked the resulting formation of ozone as the fire plumes spread downwind.

Reducing wildfire damage and promoting sustainability are two sides of the same environmental coin. As our planet faces shrinking resources and homeowners confront increased threats from climate change-related events, the housing industry must seek new systems and best practices to ensure a truly sustainable, renewable, and safer future. Developing and implementing solutions designed to protect homeowners, their property, and the environment from wildfires, will require a collaborative stakeholder commitment—from architects, contractors, and insurance providers to municipal officials, local firefighters, and state legislatures. In the end, achieving our goals of sustainability and wildfire protection will be driven by innovative thinking and problem solving, foresight and insight, purpose and passion.

I can trace my personal passion for fire prevention back to my days as a carpenter at Local 940 in Brooklyn. But it was so cold in the winter that I relocated to California in 1975 and worked doing framing during the big housing booms. From there I started working as a technical engineering representative in the lumber industry, promoting what I call a balanced build. Think of it like a balanced diet: you can't just eat meat; you need to have some vegetables too. We've a lot of old infrastructure that has to be replaced. So while we'll still use concrete footings in the ground, and we'll still use steel, at the same time we must use more renewable and sustainable materials and systems. We have to create a balanced build going forward because we can't continue to operate the way we essentially have since the late 1800s.

It can't be said enough: we need to support everyone involved to promote new, better, more sustainable, and resilient ways to reduce loss while we reverse the enormous damage done to this planet since the Industrial Revolution. That is how we get to a balanced and sustainable build.

To achieve the imperative of sustainability, those of us in the trenches need to help educate the public so everyone more fully understands all the different elements that must come together, the industry and materials issues that must be addressed, and the minds that must be opened to embrace a new, better way. In the next chapter I'll explain the development of engineered wood and the role it plays in our efforts for a sustainable future.

Chapter One:

WHY ENGINEERED WOOD

———

When ancient humans spent their days hunting and gathering, hunkering down in a cozy cave and sleeping in a tent made of mammoth skin and bones, it was both convenient and efficient for their nomadic lifestyle. But once agriculture developed, more permanent shelter was needed. In addition to bricks and mud, one of the earliest natural materials used for housing was wood, dating back ten thousand years to the ancient builders in Egypt and China. A few millennia later farmers in Europe were building Neolithic long houses, tube-like dwellings made of wood that could accommodate about thirty people.

For most of human history, our use of wood usually came at the expense of the environment. *National Geographic* notes: "Deforestation has greatly altered landscapes around the world. About two thousand years ago, 80 percent of Western Europe was forested; today the figure is 34 percent. In North America about half of the forests in the eastern part of the continent were cut down from the 1600s to the 1870s for timber and agriculture. China has lost great expanses of its forests over the past four thousand years and now just over 20 percent of it is forested. Much of the earth's farmland was once forests."

Deforestation creates a vicious atmospheric cycle. Trees take in carbon dioxide from the air for photosynthesis, so the fewer the trees, the more carbon dioxide that doesn't get captured. Also, in areas where forests are burned to create farmland—especially in the rain forests—setting trees on fire releases the carbon chemically locked in the wood, which returns to the atmosphere as carbon dioxide. With fewer trees around to take in the carbon dioxide, this greenhouse gas accumulates in the atmosphere and accelerates global warming.

Deforestation also leaves soil more vulnerable to erosion, which make plants more vulnerable to fire as the forest shifts from being a closed, moist environment to an open, dry one. While deforestation can be permanent, the good news is that it doesn't have to be. In North America, for example, forests in many areas are returning thanks to conservation efforts as well as new building practices and evolving materials. One of the current trends is the use of engineered wood, which like its solid wood counterpart, has been used for thousands of years.

Types of Wood

When it comes to solid wood, there are two main varieties: hardwoods and softwoods. Hardwood comes from trees that have flowers such as maple, oak, and walnut, which take a long time to grow. Softwoods come from coniferous trees like cedar, fir, and pine, which grow relatively faster than the hardwoods. As building materials, hardwood is more expensive, softwoods less so. Hardwoods' close grain make it extremely durable and low maintenance. Softwoods' fine wood texture require more care. When it comes to burning, as a rule hardwoods will burn hotter and longer; softwoods will ignite more easily so it makes excellent kindling but burns to ash much more quickly. Softwoods are the source of about 80 percent of the world's lumber.

But not all softwoods are created equal. Douglas fir trees from the Pacific Northwest produce the strongest lumber of all species grown for affordable wood-framed buildings. It is unique among all softwood species in that it is dimensionally stable without being dried, meaning that it does not shrink significantly. Today Douglas fir is widely distributed throughout the United States and is preferred by lumber yards because it does not twist like the white woods with lesser strength.

Engineered wood—also called mass timber, composite wood, man-made wood, or manufactured board—is an umbrella term for a variety of derivative wood products that are manufactured by binding or fixing strands, particles, fibers, or boards of wood with adhesives or other methods to form composite material.

Archaeologists have found traces of laminated wood in the tombs of the Egyptian pharaohs. In the eleventh century Chinese artisans shaved wood and glued it together to make furniture. Most engineered wood through the centuries was made from hardwood; it wasn't until the 1900s that composites made from softwood appeared. Some homeowners prefer engineered hardwood flooring over solid hardwoods because it is more resistant to changes in temperature and moisture.

The classic example of engineered wood is plywood, which is made out of thin wooden sheets that are glued together and compressed to create a block of durable wood. Bonding together cut or refashioned pieces of wood forms a composite that is stronger and stiffer than its components. Cross-laminating layers of wood veneer improves the inherent structural advantages of wood by distributing the along-the-grain strength of wood in both directions.

Even though versions of plywood had been used for centuries, it wasn't until the early twentieth century that it became a true business in its own right. In 1905 Portland, Oregon, hosted the Lewis and Clark Centennial Exposition as part of celebrating the one hundredth anniversary of the legendary expedition. Several local businesses were participating with exhibits, including the Portland Manufacturing Company, a small wooden box factory.

The plant manager decided to laminate wood panels from a variety of Pacific Northwest softwoods and displayed them as three-ply veneer work. The product got a lot of attention from fairgoers, especially some door, cabinet, and trunk manufacturers, who placed orders. By two years later in 1907, Portland Manufacturing was producing 420 panels a day and a new industry was born.

During its first fifteen years the softwood plywood industry primarily made one product: door panels. Then in 1920 a salesman from Elliott Bay Plywood in Seattle convinced car manufacturers to use plywood for their running boards. From there the market kept growing as the uses for plywood expanded. By 1929 there were seventeen plywood mills in the Pacific Northwest.

There were also setbacks. There were no industry standards, so product quality generally varied widely from mill to mill. And specifically automobile manufacturers had switched from plywood to metal running boards because the adhesive used with plywood at that time was not waterproof, so it was not particularly durable in the elements. In 1934 the Harbor Plywood Corporation developed a waterproof adhesive, which provided the opportunity to introduce plywood to many new markets. To help develop and promote new uses for plywood, the industry turned to its newly minted Douglas Fir Plywood Association, which used a 1938 law that permitted registration of industrywide trademarks, enabling plywood manufacturers to promote it as a standardized commodity rather than by individual brand names. During that same time the FHA also accepted exterior plywood, based in part on a new commercial standard that included performance tests for both interior and exterior plywood.

The plywood association promoted its product to the construction industry, which saw the advantages of using plywood for subfloors and sheathing, ceilings and walls, built-ins and siding. Plywood's growing reputation as a strong and durable construction material was cemented during World War II when it was declared an essential war material, meaning production and distribution came under government control. There were about thirty plywood mills, which produced as much as 1.8

billion square feet annually during the war years.

According to the Engineered Wood Association, "Plywood barracks sprung up everywhere. The Navy patrolled the Pacific in plywood PT boats. The Air Force flew reconnaissance missions in plywood gliders. And the Army crossed the Rhine River in plywood assault boats. There were thousands of war accessories made of plywood—from crating for machinery parts, to huts for the famed Seabees in the South Pacific, to lifeboats on hundreds of ships that kept supply lines open in the Atlantic and Pacific."

After the war the demand for plywood kept growing as the United States enjoyed a booming post-war economic boom. By the mid-1950s, there were 101 plywood mills churning out four billion square feet a year. Twenty-five years later that production number was more than sixteen billion square feet.

For close to fifty years the softwood plywood industry was based in the Pacific Northwest, primarily using the region's expansive number of Douglas fir. It wasn't just regionalism. Plywood makers did not have a way to effectively glue together veneer from softwood species grown in other regions. Eventually such a method was developed and in the early 1960s Georgia-Pacific opened the first southern pine plywood mill in Arkansas. That prompted the Douglas Fir Plywood Association to rebrand as the American Plywood Association to acknowledge the plywood industry was no longer confined to the Pacific Northwest.

Beyond Plywood

While Plywood is usually considered the original engineered wood, it's not the only one. The concept of reconstituting wood fiber to produce better-than-wood building materials has led to an engineered wood products industry featuring many products, which provide superior performance properties as well as make better use of forest resources.

In the early '80s the principle behind plywood led to the development of oriented strand board (OSB). Instead of solid sheets of veneer, OSB is made of small wood strands that are glued together in cross-laminated layers, giving it more strength. OSB is particularly suitable for load-bearing applications in construction. Now more popular than plywood, its most common uses are as sheathing in walls, flooring, and roof decking. For exterior wall applications, panels are available. Some of the advantages of OSB over other wood-based products include:

OSB can be used for both structural and non-structural applications, such as furniture frames, decorative wall paneling, shelving, packaging and crating, pallet manufacture, dry storage pallets and industrial tabletops.

OSB's bending properties are almost comparable to plywood. If the surface of OSB is extensively sanded, it can even be used as substrate for overlays for special structural applications.

Since OSB is manufactured from small diameter trees, it has a minimal negative impact on environment as compared to plywood.

Glued-laminated timber, also called glulam, is layers of dimensioned lumber bonded together with durable, moisture-resistant structural adhesives. Glulam was first used in Europe in the early 1890s. One of the first glulam structures in the United States was a research laboratory at the USDA Forest Products Laboratory in Madison, Wisconsin, which is still in service today.

The strength of glulam products gives it construction capabilities that may appeal to builders and homeowners concerned about the environment and make it ideal for applications such as engineered wood joist, more commonly known as an I-joist, which has great strength in relation to its size and weight.

Builders use I-joists in residential roof and floor construction and in commercial applications where architects specify long spans or heavy loads. With lengths of more than fifty feet and thicknesses up to 3.5 inches for individual pieces, glulam enables manufacturers to produce thicker beams to span longer distances and support heavier loads by bonding them together.

Cross-laminated timber (CLT) is a wood panel product made from gluing together layers of solid-sawn lumber cut from a single log. Each layer of boards is usually oriented perpendicular to adjacent layers and glued symmetrically on the wide faces of each board so that the outer layers have the same orientation. This gives it the strength to be used in construction as high as skyscrapers, making it a more environmentally friendly alternative to conventional materials like concrete, masonry, or steel, especially in multi-family and commercial construction. CLT is considered mass timber, which we'll talk about in much more detail in coming chapters.

Engineered Wood and the Environment

There was a time when solid wood floors were a common feature in homes. But after World War II ended, in the need to build fast to meet the demand of housing, wooden subfloors were quickly substituted with concrete ones. This created a problem for the solid wood flooring industry because their product couldn't be installed over concrete. Solid wood reacts to the dampness of concrete and enlarges and shrinks constantly, which can lead to a number of issues.

Contractors needed to find a product that matched solid wood's beauty and affordability. Many homes were simply installed with carpets or vinyl, but consumers wanted wood flooring. The answer was engineered wood floors that were topped with a layer of hardwood. Few people can tell the difference between an engineered wood floor and one made of solid wood.

As a building material, solid wood offers many environmental benefits that matter to communities across the country. It is the only major building material that is renewable and sustainable. Wood products help to increase a building's energy efficiency, minimize the energy consumed throughout the life of the product, and is better for the environment in terms of greenhouse gas emissions, air and water pollution, and other impacts. Steel and concrete respectively consume 12 percent and 20 percent more energy, emit 15 percent and 29 percent more greenhouse gases, release 10 percent and 12 percent more pollutants into the air, and generate 300 percent and 225 percent more water pollutants than wood.

The life cycle assessment (LCA) measures the energy required for a product or structure over its period of use due to raw material extraction and manufacture through distribution, use, maintenance, and disposal. LCA tools allow building professionals to compare different products, assemblies, and building designs based on their environmental impacts, so they can make more informed choices about the materials they use. According to the assessment wood also helps reduce energy consumption across the life cycle of growth, harvest, transport, manufacture, and construction compared to other structural building products.

A study from the Consortium for Research on Renewable Industrial Materials compared the construction and performance of houses in the cold climate of Minneapolis and the hot, humid climate of Atlanta. The study found that in both cases using wood significantly reduced environmental risks than steel-frame or concrete-frame alternatives, highlighting wood's benefits in embodied energy, global warming potential, air emission index, water emissions, and solid waste.

All this is to say we need wood, and we need to find ways to make what we have go further so we can keep it sustainable. So the importance of using engineered wood for flooring and other aspects of building goes beyond efficiency, cost, and appearance. It has an environmental impact, some good, some not so much. The downside is that many of the glues and binders used in engineering wood can be toxic to both humans and the environment.

The appearance of engineered wood can be aesthetically lacking so if the wood is used in a place it can be seen, the manufacturer will cover the wood with a laminate made with anything from a real wood veneer to plastic. The issue with laminates is that they often use the same toxic glues used in engineered wood. And producing engineered wood uses a lot more energy than the milling that's required for conventional wood.

On the upside some manufacturers blend recycled plant material such as rice stalks and similar components into their manufactured wood, which can be a selling point for some customers. Making engineered wood also generates a use for wood and plant scraps that were previously otherwise disposed of. But perhaps most important is that engineered wood can help sustain our forests better. For one thing wood products can use smaller trees from well-managed forests, saving old growth for future generations.

For now the United States is still able to harvest timber from its own supply, but we're exhausting our resources quickly. According to *Scientific American*, when the European settlers arrived in the New World, trees covered 820 million acres, so America must have looked like a never-ending forest. But for the colonists up and down the Atlantic coast, trees were not seen as a resource to be protected; they were a hindrance to settlement and farming, so clearing forests and exporting timber products was a win-win. It was unimaginable that trees would ever be endangered.

To promote sustainability with the diminishing reserves of remaining softwoods, we need to find alternatives. One possibility is working with materials previously regarded as unusable, such as so-called weed species to create structural particle board (SPB), a reconstituted wood panel that's suitable for structural and exterior construction and can be substituted for construction-grade plywood. SPB can be made from low density hardwoods instead of softwood supplies, giving our woodlands some much needed time to recover under careful management.

A report by Oklahoma State University notes that OSB also offers environmental advantages. "Forest products companies are utilizing raw materials more efficiently with better technologies in more environmentally friendly ways. With this approach to managing forests, engineered wood composite products such as OSB [SPB and CLT] have gained a significant role in the world market. A decline in plywood manufacturing in many countries, due to limited large log supplies and environmental concerns, will increase production of OSB in the future."

As well as reducing the amount of solid wood used and waste, engineered wood floors can also help the environment by better circulating and conducting heat, so homeowners won't need to use as much energy to keep their home warm in the winter.

While hardwood floors are durable, they have an Achilles' heel: they are vulnerable to moisture. If exposed to too much, they'll need to be replaced, a costly expense and somewhat involved process. Engineered wood flooring can withstand moisture. And even if replacement is needed, it's easy to remove the damaged planks and replace them.

Of course, not all engineered wood floors are as great for the environment as others, so it's important that the Forest Stewardship Council (FSC) has certified the product. FSC certification shows that each stage of the process involved with creating the floor, has shown to take responsibility for the environment in some way, such as the manufacturer only uses wood taken from sustainable forests, or they recycle any waste they produce and focus on environmentally friendly manufacturing processes. Choosing an engineered wood product with FSC certification will guarantee its benefits to the environment.

Engineered Wood Products' Fiery Flaw

So from a construction perspective, engineer wood products are effective and efficient. We cannot build and support our housing starts that support our economy without them. It's our only way to provide affordable housing, and it's a renewable, sustainable resource that collects carbon when the concrete and steel its replacing both create carbon. Plus our forests can no longer support construction using solid old growth lumber. But the problem is that these products have a fire design flaw that very little has been done to solve. Specifically, a major concern has arisen about wooden I-joists, which *Fire Engineering* magazine has called silent killers.

One of the largest OSB producers uses a concrete coating on the OSB that's starting to gain some market share on a minor portion of high-density projects. But is has not gone mass market, so between 2017 and 2018 fourteen large, high-density wood-framed projects were destroyed by fire. The truth is none of the current fire treatments, whether fire-rated drywall, concrete pyrolite-coated OSB, or fire-treated plywood have or will ever defend a wood-framed structure from fire during construction when only the exterior wall has fire protection and 90 percent of the inside is raw lumber, as is the case before drywall and sprinklers are added for additional fire protection.

Wood-framed buildings are not burning down because of engineered wood or today's lumber. Fires cannot consume buildings until there is horizontal framing with wood walls sheeted with OSB and floor trusses or I-joist rolled and sheeted. That creates the conditions for a fire to consume the entire building.

Fires start low and head north to survive, climbing the kiln-dried lumber studs that are sheeted with OSB and move to the floor-ceiling assembly. Gases collect and once they hit the horizontal assembly, flashover happens in less than five minutes. It's like a firebomb monster, and it won't be contained by water from firehoses. When all the raw lumber fuel starts gaining momentum and advances, there is nothing that can be done to save a building under construction. The solution is to defend 100 percent of the lumber with a fire inhibitor that breaks the back of an advancing fire before it can get to the floor-ceiling assembly.

I've developed the Mighty Fire Breaker job site spray technology, which is UL Greenguard Gold certified—meaning it is safe, even around children and in schools—and so cost-effective that there's no reason anymore to risk fire loss when building with raw lumber. That point was driven home in 2018 when two Denver construction workers were killed in a fire.

Why gamble with lives or open the door to potential litigation when there is now technology that can protect wood-framed structures. This development—adding fire protection to our renewable, sustainable resource that supports affordable housing—was long overdue because going forward it is likely engineered wood products will remain sought after by contractors and homeowners alike because of its practical use and its environmental advantages. But with wildfires increasing every year, driven by rapidly advancing climate change, we are literally feeding wood to the fire.

No wonder Mother Nature is howling.

Engineered wood products are highly sustainable as they sequester significant amounts of carbon and increase the energy efficiency of the building. In addition, engineered wood products contain a higher percentage of scraps from timber production, which would previously be either burned, left to rot, or otherwise discarded. Reusing these valuable resources reduces the amount of timber harvesting required to maintain adequate production levels, which preserves more forest area.

But more than just preserving the forest we currently have; true sustainability requires us to replace some of what we've lost through mismanagement of our natural resources. That's the goal of the reforestation movement.

Chapter Two

CLT MASS TIMBER, LABOR SHORTAGES, AND THE ENVIRONMENT

███

Cement and steel are large emitters of CO_2, yet for years their emissions have been some of the most difficult to target and reduce because no other viable, more environmentally sustainable alternative was thought to exist for large buildings. But it was hiding in plain sight. A 2014 study published in the *Journal of Sustainable Forestry* calculated that substituting wood for concrete and steel in building and bridge construction could reduce global CO_2 emissions by as much as 31 percent.

Thanks to the evolution of engineered wood described in Chapter One, the construction industry now has new options when erecting multi-story structures in the form of mass timber panels. Studies have shown that glulam beams, for example, have superior performance characteristics and result in fewer carbon dioxide emissions than steel beams.

The American Wood Council defines mass timber as a "category of framing styles characterized by the use of large, solid wood panels for wall, floor and roof construction." Also included in the definition are sculptural and non-building structures formed from solid wood panels or framing systems measuring six feet or more in width or depth.

Specifically, mass timber construction—in contrast to light-frame wood construction—is built using a category of engineered wood products typically made of large, solid wood panels, columns, or beams used for load-bearing wall, floor, and roof construction. Similar to concrete and steel, mass timber is engineered for high-strength ratings but are significantly lighter in weight. Typically formed through lamination, fasteners, or adhesives, mass timber products are thick, compressed layers of wood that create strong, structural load-bearing elements that can:

- be constructed into panelized components
- complement light-frame and hybrid options
- be an environmentally friendly substitute for carbon intensive materials and building systems.

Several products fall within the mass timber family, including cross-laminated timber (CLT), nail-laminated timber (NLT), glued-laminated timber (glulam), dowel-laminated timber (DLT), structural composite lumber, (SCL) and wood-concrete composites.

Mass timber can be produced in slabs as big as 18'x90' and up to a foot thick by layering and laminating wood boards of pine, fir, spruce, birch, ash, or beech together, so the grain of each layer faces against the grain of the adjacent layer, which is what gives the product its strength.

Cross-laminated timber was first developed in Austria in the early 1990s as a more sustainable solution for residential construction. Later, American architects began using it once they realized it could be used for bigger, taller commercial buildings—not surprisingly, one of the first areas in the country to embrace mass timber was the Pacific Northwest. Today CLT is the most familiar and widely used form of mass timber in the United States.

There's a criterion for a construction to be considered a mass timber project. If the load-bearing structure is constructed using mass timber, then the project qualifies as a mass timber project. But if it's used as anything other than the primary structural element, then the structure can't be defined as a mass timber project. Put another way; if it holds the building up, it's a mass timber project; if it's used for an ornamental purpose, it's not. Mass timber construction can be used for almost every type of application, creating well-made, cost-competitive structures, including high-end office buildings, hotels, multi-use buildings, schools and academic buildings, and multi-family residences. For example, in 2016 a seven-story, 220,000-square-foot, mixed-use building was built in Minneapolis, at the time the tallest mass timber building in the country.

Architects, engineers, and developers find it very appealing that projects using mass timber can be completed much faster and easier. The large panels can be bolted together or slipped into place with just a little welding, making it possible to complete an entire floor per week, which is a much quicker pace that concrete and steel buildings. In construction time equals money so mass timber saves money.

Mass timber's thermal qualities can also help it withstand earthquakes, an important consideration in California. Mass timber is also more fire-resistant than raw wood in light-frame constructions. It's slow-burning qualities means it chars on the outside, creating an insulating layer that protects the interior of the wood. The greater the bulk of the wood, the better it resists ignition. Through fire prevention research and design, mass timber can be engineered to last for two or three hours in a fire so people can get out safely and the fire can be extinguished.

The American Wood Council sponsored several fire tests of engineered timber, both exposed and clad with and without sprinklers. Mass timber performed well in all the scenarios, essentially self-extinguishing after three to four hours of flame exposure. Steel, by comparison, melts in about an hour under the same conditions, which threatens the structures integrity.

A mass timber structure is better at withstanding seismic activity and high winds than steel in addition to being renewable and sustainable. However, you have to make it safe, and you have to defend the carbon storage.

Mass Timber's Market Potential

According to the trade publication *GroundBreak Carolinas*, before 2011, the US didn't have a single dedicated CLT plant. Mass timber products had to be imported from Canada and Europe. A big boon for mass timber construction came in 2015 when CLT was incorporated into the National Design Specification for wood construction, leading to the 2015 International Building Code (ICC) recognizing it as a building timber product. The ICC has proposed allowing certain mass timber buildings to be up to eighteen stories high and 270 feet tall. Once the new codes are passed, there are at least seventy tall wood building projects in the works that can then get underway.

The mass timber movement has forced the industry to challenge perspectives about where wood fits in the construction industry. Even though the products have been analyzed and tested, the main issue holding up mass timber from increasing its American market share is building codes. After the catastrophic nineteenth century Chicago and San Francisco fires—which were fed by stick frame wooden structures—building codes were enacted that made wood construction of large buildings difficult.

Technology has outpaced laws, so not every state has updated its codes and regulations for mass timber product construction. In some states, codes vary from municipality to municipality so the red tape can be frustrating. Then there's also the misinformation campaign. We need builders, their investors, framers, lumber providers, and producers to defend wood-framed buildings from the negative statements claiming lumber-built buildings are cheap and unsafe.

After Portland, Oregon, passed an ordinance permitting new development geared toward seniors to be built with cross-laminated timber, a coalition of the National Ready Mixed Concrete Association issued the following statement:

"Our parents, grandparents, and great-grandparents deserve the peace of mind that comes with a stable and resilient home, and that means noncombustible building materials like steel and concrete. Unfortunately, certain developers in Portland have made it clear they would rather tout questionable environmental factors over security as the justification to use this wooden building material, leaving some of our most vulnerable members of the community at risk."

What's important to remember is that if a Class A fire protection was used on the CLT, then those buildings are better for everyone.

All American companies that build high-density housing with lumber need to help us to defend the reputation of lumber—our great renewable, sustainable building material that collects carbon to clean our air from pollution—from stories on the web that are pumping out misinformation to negatively impact the public's perception of wood buildings in order to benefit the concrete and steel industries.

But even with the ongoing efforts to discredit wood, the interest in mass timber shows no sign of slowing, in large part because it's also a renewable resource that can reduce a building's carbon footprint. Research found that using wood products for construction projects could save not only on global CO_2 emissions but also from 12 to 19 percent of global fossil fuel consumption.

Timber is 100 percent solar-powered, 100 percent renewable, and 100 percent recyclable. So we need to give builders every incentive possible to move away from steel and concrete to wood.

In 2018 Congress modified 45Q, a performance-based tax credit incentivizing carbon capture and sequestration. It's similar in function to the production tax credit (PTC) for wind power under 45Q, that enables power companies and industrial facilities to generate a tax liability offset per captured tonne (2,462 pounds) of carbon dioxide. This tax credit could change the way we look at building—including even high-rise buildings—with mass timber CLT instead of steel.

Also in 2018 Mighty Fire Breaker put together a proposal for a carbon credit project related to IRS Section 45Q in an effort to show lawmakers how it could really kickstart builders to consider moving more toward mass timber verses steel, even for high-rise buildings. The incentive to calculate a carbon credit value on all wood-framed buildings is based on our ability to fire -defend the carbon sequestered in the timber so it's never released. Our Mighty Fire Breaker carbon project is supported by clean fire protection that will defend the carbon that is stored in lumber. If the carbon project helps to reward builders with a carbon credit tax credit for defending lumber that has carbon storage, that will support the green movement better than the building industry has ever seen while protecting future generations' air quality.

The Section 45Q tax credit was originally designed for power plants when first enacted in 2008. But now our focus is on climate change and sustainability. We know trees, whether in an urban setting or in reforestation project factories, sequester tons of carbon dioxide from the air and make oxygen. This carbon sequestering process provides the support trees need to grow strong fiber, which we can use to support our best renewable, sustainable resource; lumber helps us better meet the demands for affordable housing in a carbon-neutral way.

We also need more construction workers to support the companies that are trying to raise the bar in defending wood-framed buildings from fire during construction. We also need to advocate for innovative ways to stop the advance of wildfires, which churn out greenhouse gases and threaten property and lives. Our government needs to understand that if they reward builders that use only lumber, then our reforestation factories would need to plant more acreage, which is a good thing. It should be clear that the mass timber movement and tree factory reforestation programs are part of the same sustainable chain because the green engineering of planting perfect saplings is like adding wind and solar energy to reduce our dependency on fossil fuels, which contribute to greenhouse gases.

Carbon Capture and the Sustainability Goal

If we want to create a sustainable planet, each industry needs to figure out how to play its role. And we need to accelerate best practices and technologies to help offset the carbon produced by population growth. In the building industry we can start by embracing wood in general and mass timber specifically.

Using wood helps to sustain our forests and increases carbon storage, making it the most environmentally friendly material available for building residential or commercial structures. More wood is grown each year in the United States than is harvested. None of what's cut down is wasted when trees are used to make wood products. In addition to the raw lumber, bark, trims, and even sawdust are used as energy sources to help power wood production facilities.

The durability of wood products gives structures a long lifespan, and at the end of their initial purpose, wood products are easily recycled for other uses. Numerous studies around the world have found that wood outperforms other products when you take its complete life cycle into consideration.

The Binational Softwood Lumber Council reported:

The Consortium for Research on Renewable Industrial Materials compared the environmental impacts of homes framed with wood and steel in Minneapolis and with wood and concrete in Atlanta, the framing types most common to each city. According to the report, the homes framed in steel and concrete would require 17 and 16 percent more energy respectively (from extraction through maintenance) than their wood-framed counterparts....It also outperforms steel and concrete when compared using life cycle assessment (LCA) methodology, an internationally recognized approach to evaluating materials, assemblies, and even whole structures over the course of their entire lives, based on measurable indicators of environmental impact. Using this method, study after study has shown that wood is better for the environment than steel or concrete in terms of global warming potential, resource use, embodied energy, and air pollution.

The Southern Forests Products Association noted that a study of two nearly identical homes—one framed with wood, one with cold-formed steel—found that the builder's cost for the steel-framed home was 14.2 percent higher than the wood-framed home, and the steel-framing package cost—framing labor and material—was 42.4 percent higher than that of a wood-framing package. Even though cost differences can vary depending on labor markets and other factors, the end result is the same: wood is the more economical, more sustainable, and better for the planet because of its unique properties.

How Labor Shortages Impact the Mass Timber Movement

Beyond the inherent benefits of mass timber, the movement is also being driven by circumstance; specifically, national labor shortages impacting the construction industry. According to the "2020 Construction Hiring and Business Outlook" report by the Associated General Contractors of America, 81 percent of construction businesses are struggling to find qualified skilled labor. What began as a labor shortage in isolated areas when home construction bottomed out in 2011, blossomed into a crisis that was made worse by the global COVID pandemic and rippling effect it's had on safety protocols, workers' health, and supply chains.

One trade organization noted: "Due to the global pandemic, the construction industry lost 13 percent of the country's construction workforce by April 2020. But skilled labor was facing dire shortage issues before the pandemic."

The National Association of Home Builders (NAHB) reported,

"More than four out of five builders expect to face serious challenges regarding the cost and availability of labor in 2019... Just 13 percent of builders cited labor issues as an important concern in 2011, with the rate steadily rising over the ensuing years and peaking at 82 percent in each of the last three years (2017–2019)."

A survey of builders by the NAHB/Wells-Fargo Housing Index found that 83 percent of builders are hindered by shortages of framing crews and rough carpenters, 78 percent by a lack of finish carpenters, and 70 percent to 73 percent by shortages of bricklayers, masons, and concrete workers. Nearly every trade is listed in the report—including all mechanical trades, interior finishing trades, even landscapers, excavators, and weatherization workers—with more than half of builders feeling hindered by not being able to fill these key positions.

The roots of the current labor shortage can be traced back to the Great Recession in 2009. Many skilled craftsmen dropped out of the industry during that time and didn't return, and the industry continued to feel the repercussions of that more than a decade later. Then the economy recovered—but maybe a bit too well.

A construction industry consultant observed that, "Generally speaking, most people view low unemployment rates as a positive sign of a prospering economy."

However, rates in America were so low before the pandemic that it made it harder to fill positions because fewer skilled people were searching for jobs.

Some of that is attrition; boomers are retiring. According to the Bureau of Labor Statistics, about 32 percent of construction laborers in 2018 were between forty-five and sixty-four years old. And as that current crop of skilled construction workers starts retiring, there are fewer apprentices available—or interested—to take their place.

Some is constriction; fewer in the younger generations are considering construction as a viable career option, fewer high schools offer shop classes, and more parents are guiding their children to attend four-year colleges and pursue white-collar professions.

But even for those young people interested in learning construction skills, getting training can be difficult. Many companies do not invest enough—or any—time and money into ongoing training projects, so prospective laborers who need to learn and obtain the skills sought by the construction industry do not have the necessary resources to do so with ever fewer apprenticeships and internships to be had.

Apprenticeship programs typically allow the employer to see potential employees in action to determine abilities and fit before hiring them full-time, while the apprentices get the opportunity to gain valuable hands-on work experience. Without apprenticeship programs employers have to hire without getting a firsthand evaluation of their skills. If more construction firms were offering apprenticeships, it might not only attract younger workers, but it would also act to give new workers a head start on their skill sets so that they might become highly skilled construction workers later in their construction career.

Whatever the myriad reasons behind the shortage, the demand for housing is stretching the current construction workforce thin. Due to the lack of skilled laborers on job sites, it's taking more time for structures to be completed, with some projects being cancelled altogether. This shortage of labor has prompted contractors to look for ways to reduce the number of workers needed on any given project.

One way that firms try to optimize their worker needs is by offering overtime hours. While existing workers benefit financially, that's a band-aid solution. There's only so much overtime a worker can physically handle. Plus paying employees overtime on a regular basis is not financially sustainable for most companies. Other construction employers hire lesser-skilled workers, a risky option because adhering to safety regulations is crucial. Even non-life-threatening mistakes on a job site can cost a firm thousands of dollars and a loss of reputation.

Ironically, this labor shortage could be a boon for the environment because there is another solution: using mass timber. Because the panels that use mass timber are manufactured off-site, it shortens construction time, which in turn reduces on-site skilled labor needs.

As a trade organization concludes: "With labor shortages impacting much of the construction industry, the reduced labor requirements associated with mass timber can be a significant advantage."

Our country's construction industry may be swamped with housing starts—which are the foundation to our economy—with spending on new projects peaking at well over $1 trillion, but all the housing starts and all the dollars in the world can't fix the industry's persistent labor shortages if we continue to build the same old way on-site. So to meet our nationwide need, builders are changing direction to now use off-site, factory-built prefab products—modular and panelized systems—that require less labor on-site and are more efficient. We're also seeing automated plants showing up offshore. We still have a long way to go before mass timber becomes the building material of choice, but it's already clear that such technology can make a significant dent in solving the skilled labor shortage both practically and as a matter of perception for our future workforce.

Remodeling magazine stated: "Young people today don't want to align with outmoded, underserved, and culturally marginalized social groups. Put bluntly: young folks don't want to be associated with what they perceive as low-class work. Where the various sectors of the construction industry do align with positive media and technologically sophisticated challenges, young participants are showing a high degree of interest, even showing great innovation and leadership."

The McKinsey Global Institute, a think-tank that has surveyed construction progress worldwide asserts that modular and off-site building solutions that lean on robotic and other highly technological processes will become a talent draw. The organization also found "the construction industry is not advertised to millennials nearly as much as it should be. Even in schools, there is a lack of occupational education; students learn many different subjects, but rarely do they learn their application to the real world."

The construction industry can admittedly be set in its ways, so it's been typically slow to adopt new technology to optimize construction schedules and increase efficiency. Mass timber is just the first step. As digital technology advances, we'll need fewer but even more highly skilled workers. Smart technologies already available—from drones to robots programmed for bricklaying and rebar tying—reduce costs, speed up project schedules, and lessen the impact of the labor shortage. At the same time, modernizing construction could reinvigorate its appeal as an exciting field that can provide young workers with rewarding—and lucrative—careers.

A Word about American-Grown

As our building industry tries to overcome the labor shortage with a push toward factory-built prefab walls and mass timber CLT, it's time for prefab shops to begin looking harder at the value of using stronger American-grown lumber.

American-grown Douglas fir is stronger than Canadian Spruce, and it's time our builders and engineers start building with it. They should have been comparing the design values of Douglas fir and Southern Yellow pine long before the US Government imposed tariffs on lumber imports from Canada. The Commerce Department contended that Canadian companies were selling lumber into the United States at unfair, subsidized prices, affecting builders that are not proactive in value engineering.

Canadian mills have lined our borders for years, using prefab shops and truss plants to ship across the border to avoid tariffs. If American prefab plants turn their design values toward US lumber values, tariffs will have no impact on them or housing costs whatsoever. As a country we should always look to support American-grown and American-made instead of just offshore, cheaper imports. We should promote American pride by using American-grown lumber, and design with Douglas fir and Southern Yellow pine instead of Spruce Pine fir.

But whatever wood we use—or grow—we must fire-defend it using the best technology at our disposal.

Chapter Three:
THE WIN-WIN VALUE OF REFORESTATION

———

There are few things more symbolic of nature than a healthy, verdant forest. And it turns out time spent among trees is good for more than fresh air. A 2017 *Forbes* article explained that living near a forest could actually make you happier. "It's been confirmed many times that humans are better able to cope with chronic stress and are happier when connected with nature. However, a long-term study conducted by the Max Planck Institute for Human Development and published in *Nature Scientific Reports* found that forests in particular are one of the best remedies."

While trees may soothe our souls, they also play an integral part in our planet's environmental well-being on several fronts.

Strengthen the ground. The roots of trees hold the soil in place, preventing erosion or landslides.

Improve soil. Decomposed leaves that have fallen from forest trees become humus soil., which is a form of mature compost that is an extremely nutritious planting material. Also, the roots of trees disaggregate the soil and increase its depth. Over time most solid rocks and any compact soil will turn into a soft vegetable soil that's excellent for growing plants.

Help prevent flooding. Forest canopies can reduce the impact of rain so less sediment erodes from riverbanks into the rivers, which in turn affects the ability of the river to hold more water. Trees can modify soil conditions allowing more water to be absorbed into the soil instead of running off into streams and rivers. Trees also prevent freshwater lakes from losing moisture and drying up. And more than half of our drinking water is collected and filtered by our forests.

Boost crop production. It's estimated that crop production is 25–30 percent higher when there are some forests nearby, a fact true all over the world.

Moderate temperature. Transpiration is the evaporation of water from plants, mainly through the stomates of leaves, which are small openings that takes in carbon dioxide and emits oxygen during photosynthesis. During transpiration trees release water into the atmosphere through their leaves. As the water vaporizes, the surrounding air is cooled. It's a similar process to sweating in humans.

Because of tree transpiration, water evaporation, and precipitation, forests are about 15 percent more humid than unforested areas, so at a local level, forests reduce heat in summer and cold in winter. The effect is especially beneficial in urban areas where heat gets trapped by concrete and asphalt surfaces. Trees also reduce heat by shading houses and office buildings, reducing the need for air conditioning by up to 30 percent, which in turn reduces the amount of fossil fuels burned to produce electricity. Forests also lower soil temperatures in the summer by sheltering it from direct sunlight and radiations.

But perhaps the most important impact of forests is that they make the air we breathe cleaner by removing (sequestering) CO_2 from the atmosphere during photosynthesis, a chemical reaction that produces food for the plant to survive, and emits oxygen back into the atmosphere as a byproduct. That is why forests are sometimes called Mother Nature's lungs. Trees are effective air filters by design, filtering out not only gases that are harmful to humans, but also harmful to the earth's ecosystems as a whole.

But our current crop of global forests cannot keep up with manmade air pollution. Our modern world—cars, factories, agriculture, construction—spews out CO_2 and other gases that trap heat generated by the earth in our atmosphere rather than letting it release into space. The result is what scientists call a greenhouse effect. About half of the greenhouse effect is caused by CO_2. The other pollution culprits include:

- Sulfur dioxide (SO_2), which primarily comes from burning coal for electricity and home heating (60 percent), refining, and the combustion of petroleum products (21 percent).

- Ozone (O_3), a naturally occurring oxidant that exists in the upper atmosphere. Automobile emissions and industrial emissions released in the air undergo photochemical reactions in sunlight that release ozone. Small amounts of O_3 can be formed by lightning.

- Nitrogen oxides (NOx) are primarily produced by automotive exhaust, formed by high temperature combustion when nitrogen and oxygen are present—two natural gases on earth.

- Particulates—which don't get as much press as the gases but can be just as environmentally problematic and unhealthy—are small particles emitted in smoke from burning fuel, especially diesel, which we breathe in and can cause respiratory problems. In areas with ample trees, studies show there is up to a 60 percent reduction in street-level particulates. Forests can significantly improve public health by catching dust, ash, pollen, and smoke on their leaves, keeping it out of our lungs.

Because of their ability to digest these harmful gases and particulates, trees act as so-called *carbon sinks* that help alleviate the greenhouse effect. It's estimated that each acre of new forest can sequester about 2.5 tons of carbon a year. Young trees can absorb thirteen pounds of CO_2 annually, reaching their peak CO_2 sequestration when they are about ten years old and absorb almost fifty pounds of CO_2 over the course of a year. At the same time, they are releasing enough oxygen back into the atmosphere to support two people. When you add CO_2 sequestering, carbon storage in wood, and their cooling effect, it all adds up to make trees extremely efficient tools in fighting the greenhouse effect.

But a tree's ability to sequester carbon has a downside. When cut down or burned, the carbon captured during the lifetime of the tree, doesn't just disappear. It is released back into the air as carbon dioxide. If trees are left to rot, either from logging, fire, or death, then methane—a greenhouse gas that's 30x more potent than carbon dioxide—will be released into the atmosphere during decomposition. Careless deforestation globally accounts for 15 percent of emissions worldwide of methane and CO_2, a dangerous level of these heat-trapping gases.

But we have the ability to lower that percentage. Consider this: in neighborhoods with more tree canopy cover, air quality improves by as much as 15 percent. Planting trees remains one of the most cost-effective options of removing excess CO_2 from the atmosphere. It's been said that if every American family planted just one tree, the amount of CO_2 in the atmosphere would be reduced by one billion pounds every year—to put our impact on the environment in perspective, that's just 5 percent of what human activity and industry pumps into the air we breathe each year. Over a fifty-year lifespan, a tree generates almost $32,000 worth of oxygen, providing $62,000 worth of air pollution control. This tree would also be responsible for recycling $37,500 worth of water and controlling $31,000 worth of soil erosion.

The US Forest Service estimates that combined, the forests in the United States sequestered approximately 309 million tons of carbon annually from 1952-1992, offsetting approximately 25 percent of human-caused carbon emissions during that same time. The Worldwatch Institute's *Reforesting the earth* estimated that for every ton of new-wood growth, about 1.5 tons of CO_2 are removed from the air and 1.07 tons of life-giving oxygen is produced. So think of what would happen if we increased the number of trees and forests in the world through reforestation initiatives. It's estimated that planting 100 million trees could reduce an estimated eighteen million tons of carbon per year and as a bonus would save American consumers $4 billion each year on utility bills.

Just as it sounds, reforestation is replanting trees in an area that has lost its forest cover either through fire, disease, logging, clearing for agriculture, clearing for factories and residences, or climate change. This is not a new effort. In September 1875 physician and horticulturist John Aston Warder founded the American Forestry Association. (AFA) was renamed American Forests in 1992 to reflect its conservation focus.) Born in 1812, the eldest son of Quakers parents, he developed a love of nature early in life, spending great amounts of time in the Pennsylvania woods near the family's suburban home. After earning a medical degree in college, Warder worked as a doctor for twenty years, but nature remained his passion, so he quit medicine and became involved with various horticultural associations. During the 1860s Warder's primary interest shifted to forestry, and he was one of the first to advocate for planting clusters of trees on the plains in the western United States to both provide shelter from wind and protect the soil from erosion.

The AFA founded Arbor Day in 1882 to promote planting more trees to protect and maintain America's forests. Today, the American Forests' website notes: "Since the first Arbor Day, convened by our organization in 1882, the simple power of planting more trees has provided a centering strategy—one way we can always advance America's forests."

Warder died in 1883 but his organization continued. The Civilian Conservation Corps (CCC) was created in 1933 by Franklin D. Roosevelt as a key component of the president's New Deal, driven in part because of the AFA's successful tree planting initiatives to aid recovery from World War I. CCC workers planted three billion trees to heal degraded landscapes and to create millions of jobs during the Great Depression.

We are in desperate need to continue that dedication to reforestation. Climate change, drought, insect infestation, and wildfire are devastating forests and throwing our global ecosystem out of whack.

American Forests reports: "Millions of acres of forest in the United States have been lost to human uses including development, agriculture, and mining. In Appalachia, there are more than one million acres of former surface mines that have been reclaimed to shrub, pasture, or barren land. While the sites have been reclaimed, the resulting land has compacted, rocky soils and is often covered with non-native grasses and trees. These landscapes are so inhospitable that they are stuck in their current form and do not naturally regrow into native forests."

Planting more forests can help get us back on track. I believe only a national commitment of planting billions of trees will deliver the change that's needed, not just for clean air but for long-term overall sustainability. That will include thinning overly dense forests to lower the risk of huge fires and reforesting areas that are not naturally regrowing.

In addition, we need to incorporate afforestation. Where reforestation is the replantation of trees in deforested land, afforestation is planting trees on land that has not recently been used to grow a crop of trees—usually abandoned or degraded agricultural lands. Again, this is not a new concept. The Chinese have been doing this for more than two thousand years. In the twentieth century afforestation has been a common practice in many areas of the world, including the southern United States.

While afforestation can restore previously forested areas and help to remove carbon dioxide and increase air quality, afforestation needs to be done judiciously and only after careful planning or it can hurt ecosystems and negatively impact diversity. For example, foresting savannas and other grasslands can remove specialized habitats for many animals, reduce the local biodiversity of grasses, and may introduce an invasion of non-native species.

That same level of care is also required of reforestation. Reforestation is a scientific process that must factor in more than just tree planting. Climate, landforms, and soil type also must be considered. And you have to select the right trees that will best thrive is a particular area to successfully renew forest cover, that includes woodlands and natural parks that have been decimated by reckless deforestation. Technology helps. For example, seedlings of southwestern white pine, which are threatened by white pine blister rust, are being raised by scientists, and when big enough the seedlings will be moved outdoors to start a seed orchard that will provide disease-resistant seeds for future reforestation efforts.

Urbanization

If there is an overall factor in the destruction of forests it would be urbanization, which began in earnest during the early 1800s with the rise of the Industrial Revolution, as people moved away from villages and farms to cities where jobs were plentiful. Trees were chopped down and replaced with roads, steel bridges, and railways. There was no thought to plant new saplings to make up the loss. The lack of trees combined with the smoke and particulates released from factories—and later cars—polluted the urban air.

So reforestation out in the country and other rural areas isn't enough. We also need to improve tree cover in our cities where it's even more important to improve the air we breathe.

Urban Forests

An urban forest includes all the trees that grow within an urban area:

in yards, in parks, in recreational areas, along streets, and by watersheds. In many places urban forests can be the most extensive, functional, and visible form of green infrastructure. It's estimated that urban and community forests produce $18.3 billion in value nationally by removing air pollution, reducing energy use, sequestering carbon, and avoiding emissions. In studying one urban park, researchers found the tree cover removed forty-eight pounds of particulates, nine pounds of nitrogen dioxide, six pounds of sulfur dioxide, a half-pound of carbon monoxide, and one hundred pounds of carbon *every day*. But urban forests can face unique challenges, including:

Difficult growing conditions including exposure to pollutants, heat, drought, flood, and limited space that can inhibit proper root growth and increase vulnerability to insects and disease.

Insufficient resources. Proper maintenance is essential for a thriving urban forest, which requires a commitment from the city government that doesn't always have money to spare. It's incumbent on everyone to pitch in to keep the local trees healthy and well-tended.

Developers. Show me a beautiful plot of trees and somewhere close is some developer who wants to cut everything down to build a mini mall or apartment building. Development is one of the biggest threats to our urban forest canopy and green spaces. It doesn't help that the public—and public officials—are under-informed about the benefits an urban forest provides.

Historian Jill Jones told the *Guardian* newspaper that for a long time many city officials dismissed trees as little more than expensive ornaments. But now we know the massive ecological benefits that trees provide. And as an added benefit, trees on average increase the property values by 20 percent. Ever notice the most expensive houses tend to be in areas with the most trees?

Between 2008–2016, New York planted one million trees in its five boroughs. The city's park department determined the economic impact of its trees was $120 million a year: $28 million in energy savings, $5 million of air quality improvements, and $36 million saved in mitigating storm water flooding. More than that, taking a walk down a tree-lined city street has been shown to lower a person's cortisol levels, reducing stress. Research suggests trees improve overall quality of life as people are less violent and experience less mental fatigue when they live near trees.

Geoffrey Donovan, a research forester with the US Forest Service says trees can literally be a matter of life and death. "I looked at the impact of trees on birth outcomes and found that mothers with more trees within fifty yards of their homes are less likely to have underweight babies." He also looked at mortality rates in areas that had lost millions of trees to the emerald ash borer and discovered "a corresponding increase in human mortality."

And considering that it's estimated that more than 65 percent of people on earth will be living in the cities by 2050, urban forests will be more important than ever.

So when you consider how important trees are to so many facets of our lives, to the air we breathe, and to our planet's overall environment by removing CO_2 and pumping out oxygen, it's easy to see why we need to be bullish on reforestation. In 2020 more than two dozen companies, cities, and organizations made a pledge to help plant one trillion trees around the world.

The United States' chapter—led by American Forests—has a goal of planting, restoring, and conserving more than 855 million trees by 2030. As of February 2021, thirty-five billion trees had been planted as part of the initiative. The effort was launched at the World Economic Forum in Davos, Switzerland, and was based on research by ecologist Thomas Crowther that suggested planting more than one trillion trees could wipe out years of carbon emissions while restoring forest ecosystems.

Of course, reforestation is only one part of the solution. Some people will focus on clean energy or reducing plastic use or getting rid of combustible engines in favor of electric cars. My focus is on promoting more sustainable building practices and using my company's chemical technology to control wildfires and protect homes and property, both of which promote sustainability.

Planting a trillion trees is good. Trees in the first ten years of their life sequester more CO_2 than they do over the next twenty years before they're harvested. But until the people planting understand *where* trees would have the biggest impact, the job is only half done. Don't just plant trees in parks. If we plant trees in the Pacific Northwest where there are reforestation programs that support our demands for affordable housing, it will have a real, positive impact.

As natural forests age there may be an increase in root rot as fungi travel. In our reforestation factory programs, which grow trees faster than natural forests, the saplings are separated to eliminate the rot. The better Douglas fir forests are found where climates with precipitation exceeds fifty inches annually, so our Pacific Northwest is the perfect location for its rapid growth.

There is a double benefit for concentrating on the Pacific Northwest's Douglas fir. These trees produce the strongest lumber of all species grown, for the best and affordable wood-framed buildings. Not only are Douglas firs stronger; they are the most resistant to fire even when raw. Douglas firs are unique among all softwood species because they are dimensionally stable without being dried, meaning they don't shrink or twist significantly. Lumber yards like Douglas fir better because it does not twist like the white woods that also have lesser strength, so even though Douglases are grown in the Northwest, their wood is widely distributed throughout the United States, benefiting builders and environments throughout the country.

Looked at together it's clear our reforestation programs are working as we reforest millions of acres with carbon collectors. Science is also catching up with ways to collect carbon elsewhere, and now we are consuming carbon out of the ocean with seaweed farms.[1] These are exciting times as science is learning how to give back even in steel fish farms that are growing mussels and oysters under the seaweed farms.

Planting trees supports wildfire recovery, improves water quality, mitigates climate change, cleans our air and so much more. As these trees grow and become healthy forests, they'll help fight global climate change and ensure our earth remains sustainable for today and for future generations. But that alone will not secure our environment. In the next chapter I'll explain how the mass timber movement is making the home construction industry more environmentally responsible.

---

[1] The oil and gas Industry through public pressure have started to develop massive carbon capture facilities and are using technology to locate methane gas leaks. They are also recognizing the positive affect it has on their bottom-line.

Chapter Four

DEFENDING CARBON STORAGE IN CONSTRUCTION LUMBER

—

While we are now acutely aware how important carbon capture and sequestration are, getting political acceptance and support is extremely difficult. The transitional change from fossil fuels to renewable green technology will remain as much a political difficulty as an environmental necessity. The chances of mitigating it without reducing CO_2 in our air are nil. But investment into carbon capture so far only represents 0.1 percent of clean energy investments.

Carbon capture and storage (CCS) refers to the process where CO_2 emissions from industrial processes—such as steel and cement production or from the burning of fossil fuels in power generation—are captured. That carbon is then transported from where it was produced, via ship or in a pipeline, and stored deep underground in geological formations. If we are to achieve the ambitions of the Paris Climate Agreement, we must do more than just increasing efforts to reduce emissions through mass timber. We also need to deploy existing and new technologies to remove carbon from the atmosphere, whether it's CCS, mass timber, or reforestation. Perhaps most important for the building industry is defending against greenhouse gas emissions by treating the wood used during the actual construction so they in essence become carbon storage banks.

Reversing climate change will not happen overnight. People will continue to roll the dice to live in woodland areas. Lightning will strike. Forests will burn. Arsonists will still target wood-built homes. But we need to do what we can to minimize the property and resulting environmental damage caused by fires. Fires during construction, especially from arson, is a problem that has flown largely under the radar to the general public, but it's become a scourge for the industry.

M-Fire Holdings has developed a job site spray technology that is UL Greenguard Gold certified because greenhouse gases aside, there is nothing more important than worker safety and nothing more costly than a complete fire loss during a construction project. Our spray is a cost-effective fire protection for wood-framed, high-density housing that is now being used from coast-to-coast in the United States to defend national builders from arson activity and fires of other origins, which can include:

Smoking: Many employees smoke on the job, including construction workers. With lots of heavy equipment and fuel in the construction area, a cigarette can easily trigger a fire.

Equipment: Plumbers' and welders' torches.

Heaters: Workers often use space heaters when working in colder temperatures. Unattended heaters or those placed near combustible items can also result in a fire.

Cooking: Construction workers often prepare meals right on the job site using items such as hot plates, coffee machines, and even power tools, which can all lead to fires.

Electrical: Fires caused by shoddy, exposed, or faulty wiring.

Vandalism: Construction sites without proper fencing, cameras, and other security measures are more likely to be the target of vandalism resulting in a fire.

More and more we are finding fire officials who see the same risk for buildings under construction using mass timber, especially before drywall and sprinklers are installed. This has been a focus of mine for years. The mass timber movement needs to consider defending the carbon stored in wood to support an impactful, renewable, sustainable approach to meet the demands for housing with wood.

Remember, a key benefit of constructing with mass timber is that it effectively sequesters carbon stored in lumber, providing economic and engineering advantages over concrete and steel in a *balanced build* approach. This means the construction industry can act to reduce atmospheric carbon dioxide by using products that ensure the carbon captured by wood will not be released back into the environment in the event of a fire.

Even though this protection costs less than 1 percent of the overall construction cost, which makes it less than half of 1 percent of the overall development cost, some architects and builders still believe that the American Society for Testing and Materials (ASTM) E119 test is enough. That test is used to evaluate the ability of a given wall, column, beam, floor, or roof assembly to withstand a standard fire test curve. This standardized fire test is one of several regulators rely on to gauge the relative performance of the assemblies and components that make up exterior walls.

Cities got serious about fire codes in the early 1900s after major deadly fires in Chicago, Baltimore, Spokane, and others that caused widespread destruction. Municipal officials devised tests to evaluate building construction in hopes of creating fireproof structures. According to a building trade organization, "The simple rating scale that ASTM E119 uses to rate performance in terms of hours of fire resistance is a convenient means of expressing relative fire resistance properties, and it has been accepted in the industry for decades. That said, the ASTM E119 test does not, nor does it purport to, draw direct correlations as to how a given construction method will perform when subjected to an actual fire."

In plain English, ASTM E119 is not enough. Also, our building industry gets confused that E119 testing is needed when assemblies will not protect any wood-framed building during construction. This means there is no assembly that will protect a wood-framed, high-density project under construction when 90 percent of the building is raw, kiln-dried lumber. Once it starts burning it will jump even a concrete wall assembly. The only way to defend wood-framed building during construction, before the drywall and sprinklers are in place, is with a fire inhibitor clinging to 100 percent of all that interior lumber.

That's the message I've shared in many closed-door meetings with mayors, fire officials, and builders in cities hit hard by arson strikes. Our building fire tracking map proves that all cities are vulnerable to these attacks, which won't be stopped by taller fences and high-def cameras. But now that we can offer a new, cost-effective way to defend wood framing as it goes vertical, firefighters have enough time to save the building from total destruction.

Our building industry must also realize that water will not put out the advance of a raw wood-framed, high-density fire once it gets to a flashover pace. Typical high-density buildings have nine-foot walls, and the fire can get to flashover mass in less than five minutes. And that is not the firefighters' fault. Here is what happens in a fire.

- Wood-framed buildings only become vulnerable to complete fire consumption as the walls, floors, and roofs get sheeted in. Framed walls without sheeting are not vulnerable to complete fire consumption.

- In order for fire to survive and accelerate, it needs fuel (i.e., lumber), a heat source, oxygen, and the carbon monoxide it's creating has to be left unobstructed.

- As the fire grows and consumes the fuel around it, it starts climbing up the wood studs, the vertical pieces of wood positioned at closely spaced intervals to form the framework of a wall that is sheeted with OSB (oriented strand board), a type of engineered wood discussed in Chapter One.

• As the fire climbs, it creates vertical momentum and accelerated heat temperatures, and gases start to fill the floor-ceiling assembly when the floor is sheeted.

• When the flames hit the horizontal sheet floor assembling and the temperature reaches 900 degrees, flashover begins. FEMA described flashover as "a thermally-driven event during which every combustible surface exposed to thermal radiation in a compartment or enclosed space rapidly and simultaneously ignites."

• Typical high-density buildings have 9' walls, and the fire can get to flashover mass in less than five minutes. Usually a fire isn't recognized and reported to the local fire department until after a flashover.

Firefighters understand flashover events in wood-framed buildings under construction better than anyone. So by the time they get on-site, they have already made the decision that they aren't going in and can only defend the building with boom cranes to protect the neighboring buildings and houses.

Turnkey framing contractors are especially at risk of liability if a fire consumes a building. There was a time when the builder controlled and provided the whole package:; lot development, plans, materials, and labor. In the 1940s and 1950s when demand for housing skyrocketed, so did homebuilding. Contractors looked for ways to build homes faster for less money. So the one-stop-shop method—where a division of responsibilities among the trades and outsourcing of materials and labor—gained increasing favor.

Initially the specialization seemed like an advancement because it meant every aspect of a building was built by experts in their field rather than jacks of all trade. But as light-frame construction has become more complex, communication among the disparate trade workers could break down, leading to mistakes and delays. So the tide turned again, and now the trend is looking for ways to minimize supply-side fragmentation, reduce waste, and limit mistakes. Turnkey framing gives the builder a single source to work with to get the project completed with better control over materials and productivity.

But that's also what puts the turnkey framer at an increased liability should a fire occur, especially if it hurts or takes any worker's life. The framing contractor can mitigate risk by suggesting fire protection on 100 percent of the lumber. If the builder refuses to pay the cost to defend the wood structure as it's being framed, the contractor should make the builder sign a waiver that they refused to pay for making the job site safer. The same goes for smaller contractors, who should advocate for fire protection to a homeowner, even on a room addition.

Obviously, having clean, *safe*, fire inhibitor chemistry that can be sprayed on all lumber and engineered wood products as framers erect the support structure changes the game.

Our fire inhibitor is night and day when compared to the traditional method of pressure-treated lumber, which refers to wood that's undergone a preservation process that removes air from the wood and then injects pressurized chemicals, usually fungicides/insecticides. It can also inject fire retardants.

Now, there are practical advantages of using pressure-treated wood over natural, raw wood. First, it lasts longer. The reason so many construction companies and product manufacturers choose pressure-treated wood is because it lasts longer than traditional wood. Think of how the shelf life of food is increased by adding all sorts of chemicals. The same is true with wood. The chemicals prevent natural decay and kill microorganisms that can cause rot.

But there are also some important problems with pressure-treating wood. First and foremost, it becomes toxic when burned. If a house catches on fire, even if the wood just smolders, it releases a variety of dangerous chemicals and pollutants into the atmosphere—and into the lungs of anyone near the fire.

One of the most common types of pressure treatments called CCA—which was developed back in the 1930s—contained a solution of copper, arsenic, and chromium. It took until 2002 and a report by the Consumer Product Safety Commission (CPSC) in 2002, before the Environmental Protection Agency (EPA) determined that exposure to CCA-treated wood could cause illness and began restricting its use for residential construction although it's still allowed for commercial construction applications. Beyond CCA, pressurized wood still uses other hazardous insecticide and fungicide chemicals such as ammoniacal copper zinc arsenate (ACZA). When burned ACZA increases the risk of chronic respiratory disease and even cancer.

Plus, the old school methods of pressure-treating lumber made the cost double, so it's only used primarily on exterior walls in some buildings. Those fire-treated walls do *nothing* to defend a building during construction from complete fire consumption because 90 percent is raw fuel that will rip right through a fire-treated wall. Our new school method is to spray our job site fire inhibitor on 100 percent of the interior lumber at a quarter of the cost. This new technology is equal to or better than pressure-treated lumber in its E84 extended test results, and the way it works is by breaking fire's chain reaction.

Fire typically results from the chemical reaction between oxygen in the atmosphere and some sort of fuel that is heated to its ignition temperature. Wood is an excellent fuel. The heat needed to ignite that fuel could be lightning, a spark from welding, or an arsonist's match.

When the wood reaches about 300 degrees Fahrenheit it ignites and releases:

- volatile gases, including hydrogen, CO_2, and oxygen, which we see as smoke;
- free radicals, which are uncharged, very reactive, and short-

lived molecules that are also produced during a fire; and

• char, which is nearly pure carbon.

The charcoal sold for backyard barbecues is essentially char; wood that has been commercially heated to remove nearly all the volatile gases to leave behind just the carbon. That's why charcoal burns with no smoke. Fire also produces ash, which is comprised of the unburnable minerals in the wood, such as calcium and potassium.

So when wood burns—whether on a construction site, a house fire, or a wildfire—there are two reactions going on simultaneously. First, the carbon in the char combines with oxygen, which is a slow reaction. Think of how long charcoal in a barbecue can stay hot. Second, the reaction causing the fuel to ignite and burn involves free radicals combining with oxygen in the air. Together those two reactions generate massive heat, which ignites even more fuel. Fire feeds fire, meaning that essentially the chemical reactions in a fire are self-perpetuating. They won't stop until they run out of oxygen or fuel—or something makes it stop.

We've made a product that can stop a fire in its tracks by focusing on the role that the free radical hydrocarbon molecules H+, OH-, and O- play in sustaining the combustion phase of any fire. Our technology works by capturing those free radicals, breaking the chemical reaction, and suppressing the fire's strength and ability to advance. I could walk into a pool on fire with diesel fuel and with a little canister of our product, put it out, and not worry about it reigniting around me. Our system is the first simple, effective method to make wood-framed buildings safer for occupants, workers, and firefighters. Our system also comes with an app that notifies local firefighters what properties have been defended during construction.

After several high-profile construction site fires since 2018, contractors are coming to realize that tall fences and a security guard or two are no match for an arsonist determined to burn a structure. In 2018 what was to become a five-story, eighty-four-unit apartment building in Denver was destroyed by a massive fire. More than fifty construction workers were at the site when flames shot two hundred feet into the sky.

According to reports: "The fire spread rapidly due to open ventilation and the wood-frame construction that did not have active fire stops. Some corridors had drywall, but many rooms were separated by plywood. Doors and windows had not been installed."

Firefighters reported that the flames produced a tremendous amount of heat—so hot that it was detectable from space—because there was nothing to protect the wood in the building from the fire. Some of the workers jumped from the second and third floors of the building to escape. The inferno damaged the roofs of several other buildings, destroyed more than thirty cars, injured six including a firefighter, and killed two workers.

A month later, the fire was still under investigation. "After conducting two hundred interviews and reviewing five hundred photographs, fire investigators have ruled out actions by the following workers as causes: framers, welders, plumbers, drywall installers, insulation workers, and electricians. Electrical cords and power units set up to provide a temporary power source for the construction site also have been ruled out."

There is still no official cause. Nor were Denver investigators able to identify an official cause at a second fire that happened a few months later at a second construction site that destroyed two townhome buildings under construction near the Denver Broncos' stadium. Fortunately, nobody was injured.

But a huge fire that reduced a 206-unit Las Vegas luxury apartment complex under construction to ashes in January 2021 was identified as arson. Investigators from the Bureau of Alcohol, Tobacco, Firearms, and Explosives found evidence that a flammable substance was used to fuel the midnight blaze that destroyed the luxury apartment complex southwest of the Las Vegas Strip. The project was a complete loss, with damage was estimated at up to $35 million.

A University of Arizona student housing building under construction sustained major damage from a fire so hot that it melted two construction cranes and forced the evacuation of thirty residents living in a nearby apartment complex. Damages exceeded $1 million.

Between 2017 and 2018 there were nine construction site fires in Oakland, California, which federal authorities attributed to arsonists targeting new housing developments. But Oakland was hardly unique. According to the United States Fire Administration (USFA), there are more than 4,800 construction site fires every year, causing approximately $35 million in property loss.

The agency has published suggestions for contractors to try and minimize their exposure to fires, including:

- Store solvents, fuels, and tools in a locked storage container or remove them from the job site when you are not using them.
- Request additional patrols or drive-bys from your local law enforcement.
- Remove trash and debris from the job site.
- Try not to store excess materials on the job site.
- Secure doors and windows on structures when crews are not actively working on the property.

As one legal expert explains, "Construction site fires are ultimately the responsibility of the property owner or the employer. Property owners and construction companies are legally obligated to mitigate fire hazards. Because the causes of fires are largely known, almost any construction site fire is preventable."

The reason for the epidemic of arson is unclear. Some experts think it could be anger at development and/or gentrification. Displacement is becoming a divisive issue in cities across the US, where the pressure for urban living is accelerating. The cities that seem the most targeted are the ones attracting new businesses, highly skilled workers, and large corporations, all of which drive up both the demand for and cost of housing. Subsequently, local residents—and neighborhood renters in particular—are displaced, forced to move to more affordable locations. Frustration and resentment could be part of the reason we are seeing so many high-density wood-framed projects burned down during construction over the past years.

But whatever the arsonists motivations, builders of high-density, wood-framed buildings need to consider a new way of making buildings safer for the workers. They need to use our proven, clean, affordable, cost-saving site spray technology. If they don't, they are subjecting workers, firefighters, and neighborhoods to undue risk and making themselves vulnerable to lawsuits from investors. That's why today some builders have mandated the use of our new fire inhibitor on all their wood-framed buildings going forward. Building any wood-framed building today without adding fire protection when it finally affordable is no longer worth taking to risk.

Remember, defending carbon also defends the mass timber movement. When a builder defends their lumber, they also defend the building industry that uses this renewable, sustainable building product. If you ignore that wood needs fire protection, you're essentially supporting the lobbyists who are trying to change the codes so that any structure more than two stories high must be built with concrete and steel. Future generations' air quality is dependent on us to build with lumber, which is a carbon collector versus carbon emitters like concrete and steel.

Nature Is Screaming; Are We Ready to Listen?

Steve in Washington, DC, teaching risk reduction.

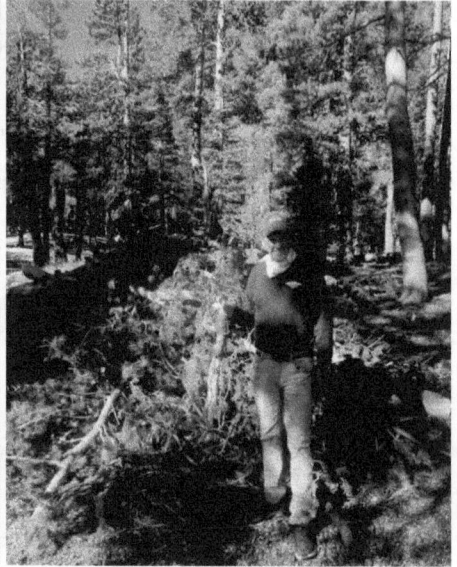

Steve is now fully committed to reducing the risk of loss during wildfire events even in the forest

Construction Safety Day in Seattle

Steve at the New York Fire Academy.

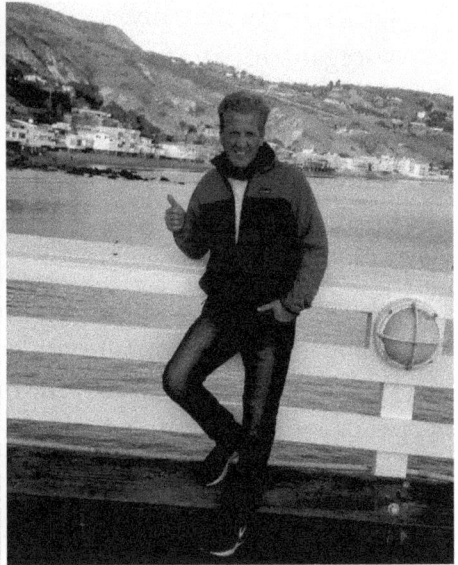

Steve is nationally recognized as the guy who brought the most cost-effective fire protection to high-density housing, defending five-story wood structures from arson and jobsite mistakes during construction.

Steve has opened the first of its kind store, the Wildfire Depot in Sonoma County, to help educate property owners about proactive wildfire defense.

Chapter Five

EDUCATING INSURANCE UNDERWRITERS ABOUT REDUCING WILDFIRE RISK

——

It could be argued that Benjamin Franklin was our country's first fire safety advocate. The same practical, inventive man from Philadelphia who helped found the United States of America made many important contributions to fire prevention during America's colonial days and early years as a nation. His contributions included:

- writing frequently on fire prevention;
- founding the first volunteer fire company in 1736 called Union Fire Company;
- organizing the first fire insurance company in 1752, named the Philadelphia Contribution-ship for the Insurance of Houses from Loss by Fire;
- inventing the lighting rod in 1753 to protect houses from burning down and people getting electrocuted from lightning; and
- inventing the Franklin Stove in 1741 to reduce the risk of fire to houses from sparks and cinders from conventional fireplaces.

Ben Franklin believed in preventing problems before they happened, which is why he told fire-threatened Philadelphians in 1736 that an ounce of prevention was worth more than a pound of cure.

Clearly, preventing fires is better than fighting them, so we feel a kinship with Ben Franklin and his efforts to defend against fire. His spirit of brotherly love motivated me to help develop the best possible technology for protecting raw lumber buildings during construction on job sites to advance the field of fire prevention. I hope Franklin would approve of what M-Fire is doing today as we follow his principles of fire prevention to advance safer and better best practices in the wood-framed and mass timber building industries.

Even back in 1752 Franklin also understood that future insurance companies would play an important role in fire prevention by setting higher standards and requiring better practices as advances in applied science would enable. Today these companies continue to underwrite new construction and building insurance policies against loss by fire, often at astronomical premium prices.

Californian's state insurance commissioner noted that people living in wildfire-prone areas without homeowners insurance have "nightmarish fears of losing everything that is precious."

"Some California insurers see wildfire-prone areas as too risky and would rather not cover certain homeowners than give them a policy that doesn't account for the greater dangers—and costs—of property damage from wildfires," Mark Sektnan, vice president of the American Property Casualty Insurance Association, told CBS MoneyWatch.

It doesn't have to be that way. The insurance industry has to stop cancelling property owner policies now and get behind new American Accredited Innovation that lower their risk of loss so property owners and their lenders are not threatened by cancellations. It doesn't have to be that way, but it is going to take educating builders and insurance companies alike.

In 2017 the *Wall Street Journal* ran a headline: Apartment Fires Are Tied to Cheaper, Wood-Based Construction. "Developers often prefer wood over steel because it is quicker to erect and can cut framing costs by at least 10 percent to 20 percent....But while fire safety officials say wood-frame buildings generally are safe once they are completed, they are particularly vulnerable to blazes before they are outfitted with their walls and safety features such as sprinklers."

While on its face this may be accurate, it's out of context. It's not younger juvenile lumber and engineered wood products that cause complete fire failure. Buildings are being lit on fire and burning so quickly because the insurance industry allows builders to build with kiln-dried raw lumber.

Builders today no longer need to take the risk to build with raw lumber. The industry is evolving and now there is a safer and more effective fire protection that can proactively defend buildings from fire coast-to-coast. It's no longer about being unable to afford building an entire structure with traditional pressure-treated fire protected lumber, which doubles the framing cost. Taking that risk is akin to saying hard hats, steel toed shoes, and fall protection is optional.

Some builders recognize risk on their own, others need it to become the law before making their buildings safer during construction. Building any wood-framed building over two stories puts all workers at greater risk for injury or even death. Gambling to build with raw lumber is an unnecessary financial risk because it makes it easier for lawyers to win easy lawsuits because they are well aware that protection exists and could have been used. Here is how one personal injury lawyer markets their services.

There are certain situations where workers' comp benefits are not your sole remedy, and you can seek compensation through a third-party injury claim or work injury lawsuit in civil court. You may be able to seek additional burn injury compensation if:

● Your employer doesn't have workers' compensation insurance as required by law

● Willful and intentional conduct was the cause of the fire

● A third party's actions caused the incident.

Third parties could be almost any individual or entity present

at the construction site. For example, you may have grounds for a personal injury lawsuit for burn injuries caused by:
a general contractor or subcontractor
- *a vendor of supplies or materials*
- *the manufacturer of defective equipment or other parties in the supply chain*
- *the owner, operator, or other entity in control of the property*
- *an engineer, architect, or designer*

Through a third-party liability claim for burn injuries, you can seek compensation for your:
- *medical bills*
- *lost wages and benefits*
- *physical pain and suffering*
- *emotional distress and mental anguish*
- *disfigurement*
- *disability*
- *reduced earning capacity*
- *reduced quality of life*

But even if a contractor never experiences a fire on their job sites, they are still financially impacted, getting slammed by high insurance premiums. Although you may find your project insurable, these construction site fire events are causing hundreds of millions of dollars in property loss, lengthy investigations, and costly revenue setbacks and rebuilds. The new sleeping giant in the building industry is no longer mold when it comes to legal issues for contractors and large builders. It's fire. And if any workers get hurt due to a fire, be ready for the big hammer to drop on a negligence claim.

Those in charge of the risk management side need to embrace new ways to defend wood structures from fire, whether it's a job site plumber mistake or a building lit on fire by someone being pushed out of their neighborhood by new developments. Construction companies that use our job site spray fire protection for 100 percent of the lumber should get a reduction in their risk management premiums because the technology protects not only the builder but the insurance provider as well. We are trying to help lower risk management premiums so we can keep up with the demands for housing using our renewable sustainable lumber.

Our wood-framed stick buildings will only stay competitive with concrete and steel if we defend 100 percent from fire, so we don't end up like CLT where there is no savings compared to concrete and steel because of the high premiums demanded by insurance carriers. I'm a huge supporter of CLT; however, all the producers drank the Euro Kool-Aid regarding their Class B E119 structure failure for mass timber high-rise, instead of adding fire protection to make it Class A to offset the negative fire risk perception.

Builders who are still in denial about how wood-framed, high-density, urban infill projects are under attack should be paying high premiums. The proactive builders that are doing everything they can to defend their buildings should be rewarded with premium reductions from insurance underwriters. The good news is that risk management underwriters are starting to pay attention to our treatment. They are realizing that when we fire-treat and protect 100 percent of the interior lumber instead of just exterior fire wall, it means they have less risk to insure and can considerably lower their premiums.

Another benefit of our fire defense system is that it also helps protect a city's water supply. When homes are burning and firefighters are out there for two days with the hose trying to make sure it doesn't reignite, all those toxic runoffs are going into the water table. By using our technology, instead of watering for forty-eight hours, you can spray our treated mulch over the top of a building, and it will shut the smoke down instantly, encapsulating the toxins from stormwater runoff so we don't destroy everybody's water table.

Insurance can be complex because a single commercial building project requires a range of insurance products, including property insurance, workers compensation, course-of-construction liability insurance and builders risk insurance. And the cost of each of these types of insurance depends on factors unique to each individual project.

To help reduce the likelihood of fire events, contractors can follow a number of best practices to safeguard construction sites against arson and catastrophic exposures. Anyone that has anything to do with framing with wood needs to recommend to the owners that the lumber has to get fire protected as the building frames vertically. In the world we all live in today, we can no longer think it could never happen to us. M-Fire now has the cost-effective solution that causes no delays on the job sites.

Mighty Fire Breaker is the number one company in the US supported by ASTM accredited testing, and when our program is installed and followed we are 100 times safer than just trimming or hoping private firefighters get to your house in time to spray foam. At the same time, you need a third party in insurance and a plan check to get positive results.

Understand, there is no 100 percent guarantee against defending wildfires, just as wearing seat belts doesn't prevent accidents. But if there is an accident and we are wearing our seatbelts, we significantly lower our risk of a serious injury. Likewise, if there is a wildfire nearby. Our system will protect your property better than any reactive response.

Premiums and nonrenewal rates have skyrocketed in California's fire-prone regions since 2015 as companies are loath to pay for damages wreaked by the state's increasingly devastating fires. In 2020 more than four million acres burned, twice the state's previous modern-day record. A moratorium enacted by the California insurance commissioner gave 2.1 million homeowners in the vicinity of those blazes—18 percent of all policyholders in the state—an additional year to find a new insurance company or take steps to mitigate fire risk on their property and convince their insurer to extend coverage.

In 2019, insurers dropped 235,274 policies in California, a 61 percent increase from 2018, according to data from the insurance commissioner office. Sixty-five percent of those came in areas of moderate to high fire risk, and the state's ten most fire-prone counties saw a 203 percent increase in non-renewals.

No matter the work they do to mitigate fire risk on their property, many residents in those regions are now seeing cancellations because they're considered too high-risk given how far California's fires have spread. Insurers say the market in some fire-prone regions simply isn't sustainable, and they're facing their own challenges paying up to reinsurers, the companies that insure the insurance companies.

The insurance department and consumer advocates both say the ultimate solution lies in insurance companies incentivizing homeowners to "harden" their homes against wildfires by making improvements like installing a metal roof or cutting back brush to leave space around the home.

As one homeowner says, "It's going to cause people like me to leave the state. If we can't afford to live here because if our house burns down we can't rebuild, what are we doing then?"

In October 2020, Ricardo Lara, California's insurance commissioner, acknowledged that residents of the state had "nightmarish fears of losing everything that is precious. The people of California are strong and resilient, but we cannot just continue business as usual," he said of the insurance challenge.

A retired teacher in the Sierra Nevada foothills told a local TV station that her insurance had more than tripled, $1,500 per year to more than $4,800. Her home had survived a 2018 wildfire, and she had cleared vegetation around her home to reduce its vulnerability, even though none of it translated to lower premiums because insurers are just going by ZIP code and fire risk zones.

What makes our Locked-N-Loaded systems so much better than

even the insurance industries' private fire fighters—for those that can afford to pay for it—is that the properties we defend are not counting on just in time defense. The properties we defend are never counting on water, power, or labor with handheld tools. Our structures are ready to defend 24/7. The reality is, if you wait to get proactive with fire defense technology, it could be too late because wildfire season is becoming a year-round scourge.

Chapter Six

CLIMATE CHANGE
AND WILDFIRES

—

Climate change is the defining issue of our time and we are at a defining moment. From shifting weather patterns that threaten food production, to rising sea levels that increase the risk of catastrophic flooding, the impacts of climate change are global in scope and unprecedented in scale.
~ United Nations

—

The images have become increasingly commonplace. Under blood-orange skies, waves of flames consume vast swaths of land and turn homes and buildings to ash. Data from the US National Interagency Fire Center shows the threat of wildfires has steadily grown in the United States since the 1980s. In 2018 more than fifty-five thousand fires burned more than 8.5 million acres of land, an area larger than the states of Massachusetts, Vermont, and Rhode Island combined. From 2015 to 2020, California reported ten of the most destructive wildfires in the state's history, which forced hundreds of thousands of people to evacuate their homes. In August of 2020 the Pine Gulch Fire became the largest wildfire in Colorado history. The Forest Service in Maine reported that through the first eight months of 2020 there had been more than 530 wildfires compared to just 356 in all of 2019.

Part of the reason is human encroachment. Developers are increasingly intruding into wildland areas to build housing communities, which increases the likelihood of fires, their level of devastation, and the number of people put at risk.

But perhaps the most significant causation factor is climate change, which dries out forest fuels—the organic matter that burns and spreads wildfire—making it more likely prone to ignition. Fire seasons are also running longer, stronger, and hotter. And this is not just a United States problem. From Australia to Europe to South America, countries are on fire, stretching global firefighting resources thin.

It's a vicious cycle that keeps getting bigger. Not only is the average wildfire season three and a half months longer than it was in the 1980s and 1990s, but the number of annual large fires in the West has tripled, burning twice as many acres.

Although human activities such as campfires, discarded cigarettes, and arson are mainly responsible for starting the fires, hotter weather makes forests drier, increasing the damage done. At the same time, winter snowpacks are melting about a month earlier, meaning that the forests are drier for longer periods of time.

We have the power to break the cycle and get on track toward a more sustainable future. We can either keep spending billions of dollars to fight fires, or we can work together to slow and eventually stop the greenhouse gas emissions from warming our planet.

That said, it should be understood that there has always been climate change, but prior to the Industrial Revolution, it was a function of the natural world, such as variations in sunlight caused by slight wobbles in the earth's orbit. Variations in the sun itself impacts the amount of solar energy that reaches Earth.

NASA notes, "Volcanic eruptions have generated particles that reflect sunlight, cooling the climate. Volcanic activity has also, in the deep past, increased greenhouse gases over millions of years, contributing to episodes of global warming. These natural causes are still in play today, but their influence is too small or they occur too slowly to explain the rapid warming seen in recent decades. We know this because scientists closely monitor the natural and human activities that influence climate with a fleet of satellites and surface instruments."

It's no small irony that according to research published in the journal *Science*, the earth has actually been in a natural cooling period for the last two thousand years, based on its position in relation to the sun. But that cooling has been overcome over the last several hundred years by warming that coincided with increased emissions of greenhouse gases caused by industrial and commercial use of fossil fuels.

It also needs to be understood that the greenhouse effect is not in and of itself bad for Earth. Quite the contrary. It's the natural process that warms the earth when certain gases trap heat in our atmosphere, the way a greenhouse does. Those gases—carbon dioxide (Co_2), methane, and nitrous oxide—are necessary to keep the planet from being a frozen wasteland. According to the World Meteorological Organization, without the greenhouse gas effect, Earth would be uninhabitable.

For thousands of years the earth had naturally regulated the concentration of these greenhouse gases. But the atmosphere was thrown out of whack once we began burning fossil fuels, which pumped massive quantities of CO_2. The result was a more rapidly warming planet—a phenomenon we've known about officially since the mid-1960s when a presidential advisory committee panel voiced concern about the greenhouse effect. Ten years later Columbia University professor Wallace Broecker, a geochemist, published a paper putting *global warming* in the title and into the national consciousness.

The predicted consequences of inaction have come to pass, such as the 2018 government report that painted a dire picture about wildfires. The report warned that the continued release of greenhouse gases from cars, factories, agriculture, and other sources would make wildfires more frequent. And because wildfires release a lot of CO_2 into the atmosphere, they exacerbate the climate change causing more wildfires to happen in the first place.

Another consequence is that fires are actually getting bigger and hotter, making them more deadly and harder to contain. Scientists now calls these conflagrations *megafires*, which account for an increasing proportion of the total area burned in every year. According to one report, climate scientists predict "the number of days conducive to such fires will increase by as much as fifty percent by the middle of the century."

> I turned and saw a massive fire cloud rising over the mountain to the east. The smoke and vapor boiled and expanded. It was a pyrocumulus—the first I'd seen. I took a picture with my phone, then sat and stared. Over the next hour, I watched its white, cauliflower-like head rise to twenty-five thousand feet. It looked like a mushroom cloud. I tried to imagine the combustion taking place below it—the heat and speed of a fire that could send so much smoke and ash into the sky. What powerful, nefarious force was creating this beast? It was us.
>
> ~ M. R. O'Connor

Megafires can generate a cloud of so much smoke and heat that it can turn into literal storms called pyrocumulonimbus, which NASA defines as "a storm that combines smoke and fire with the features of a violent thunderstorm," and describes as "the fire-breathing dragon of clouds," which can extend five miles into the sky. At the center of the fire, hot air rises rapidly, creating a vacuum that is filled by cooler air flowing into the fire from all directions, which then becomes more super-heated air, making the cloud bigger. When it gets big enough, that heated air can shoot out in front of the flames and cause forests to spontaneously combust even before coming into contact with any flames.

Twenty years ago, many scientists thought it wasn't possible for a wildfire to cause such a phenomenon. The *New Yorker* reported: "Today's largest fires behave in surprising ways. In the late nineteen-nineties, a few scientists began inspecting satellite images of unusual clouds over Australia and elsewhere; the meteorologist Michael Fromm speculated that they could be connected to the convective force of giant wildfires below them. Eventually, the researchers confirmed that particularly powerful wildfires could cause not just pyrocumulus clouds but vast firestorms called pyrocumulonimbus columns. Created by the flames at ground level, the columns are tall enough to generate lightning, and their air currents are so strong that they can punch particles of smoke into the stratosphere, where commercial jets typically cruise. Skeptics believed that if you saw aerosols in the stratosphere it had to be a volcano."

They were wrong. Today we know fire storms are real and they are one reason the cost of fighting fires has gone up exponentially. In 2018 the federal government spent over $3 billion fighting wildfires, and the cost goes up each year—as does the amount of CO_2 the fires produce.

Scientists at the National Center for Atmospheric Research and the University of California used satellite data about fires and a computer model to estimate the amount of CO_2 wildfires release based on the amount of vegetation burned. Their research found that fires in the contiguous United States and Alaska release about 290 million metric tons of CO_2 a year, which is about 6 percent of the greenhouse gases released in the US from burning fossil fuels.

Mother Nature Is Screaming but Was Anyone at the Global Climate Summit Listening?

At the 20211 United Nations global climate summit in Glasgow, Scotland, more than two hundred countries agreed to increase efforts in addressing the climate crisis, including new pledges regarding methane gas pollution, deforestation, and completing rules on carbon trading. For example, more than one hundred countries have now joined a coalition—led by the United States and European Union to cut current methane gas emissions 30 percent by 2030. But China, Russia, and India, which together account for more than a third of all global methane emissions, did not join the coalition.

Speakers at the summit begged governments to commit to slashing other greenhouse gas emissions and to provide more available funding for nations most vulnerable to a changing climate. The biggest producers of CO_2 stopped short of such pledges, which prompted some climate scientists to express frustration and concern that the incremental progress made in Glasgow was better than nothing but inadequate to address the severity of the climate crisis.

Ani Dasgupta, president and chief executive officer of the World Resources Institute, said, "While we are not yet on track, the progress made over the last year and at the United Nations Climate Change Conference summit offered bright spots. The real test now is whether countries accelerate their efforts and turn commitments into action."

Barry Rabe, a senior fellow at the Brookings Institution, said, "The Glasgow meetings serve as a reminder of just how hard it is to achieve transformational progress on climate change in a few weeks, despite all of the melodrama. That said, [there's] some real progress here on issues such as carbon markets, coal transition, methane and more. The emerging question is whether these areas of agreement can be implemented."

Grass Roots Efforts to Fight Climate Change

It's not just governments addressing climate change. There are numerous international NGOs that are having a positive impact on efforts to reduce greenhouse gases. For example, the Program for the Endorsement of Forest Certification (PEFC), is a leader in forest certification systems.

"As an international non-profit, non-governmental organization, we are dedicated to promoting sustainable forest management through independent third-party certification. We endorse national forest certification systems developed through multi-stakeholder processes and tailored to local priorities and conditions. We believe that forest certification needs to be local; this is why we choose to work with national organizations to advance responsible forestry. Forest certification is at our core. We provide forest owners, from the large to the small, with a tool to demonstrate their responsible practices, while empowering consumers and companies to buy sustainably."

For example, they spread the word about how plastic packaging, synthetic clothing, and the concrete and steel used in home construction all have a heavy carbon footprint. Plastic accounts for 4 percent of the global oil production, but we can make materials with similar properties to plastics from tree fibers. Using timber for construction instead of concrete could reduce global CO_2 emissions. And as previously shown, with engineered wood we can now construct buildings higher and faster. Instead of wearing clothes made from water and energy hungry synthetic fibers, we can make cloth from forest-based fibers. Products made from sustainably managed trees and other forest flora can provide a healthier alternative. Provided they come from forests, wood products help offset greenhouse gas emissions for their lifetime—and beyond. Timber continues to store carbon, even after the tree has been harvested and used. Furniture and wooden homes can store carbon for hundreds of years.

Based in Geneva, Switzerland, PEFC was established in 1999 by a group of European forest owners to promote sustainable forest management. Today PEFC is the largest forest certification system in the world.

Certification is important because forests are integral to combat climate change and its impacts. Trees act as carbon sinks, capturing and storing carbon, and removing significant volumes of CO_2 from the atmosphere. Moreover, forest products offer a carbon-neutral alternative to fossil fuels. But they must be managed sustainably to increase the amount of CO_2 that the trees remove from the atmosphere and store as carbon in their biomass.

The PEFC website notes: "Forest management also strengthens the capacity of people and communities living in and around forests to adapt to the effects of climate change. It provides benefits to all, from supporting local livelihoods and protecting biodiversity, to reducing rural poverty. Forest loss and deforestation is prevented by prohibiting exploitative activities like the conversion of forests to agricultural land, while protecting areas with high-carbon stock. Climate-positive practices in forest management are promoted, including greenhouse gas emission reductions. In addition, PEFC requirements address the environmental impacts of droughts, floods, storms, and forest fires, all of which are expected to intensify as climate change progresses."

To make sure wood in products come from a certified, sustainably managed forest, PEFC instituted a chain of custody certification, which tracks the wood from certified forests to the final product, monitoring each step of the supply chain. Finding additional financial value in forests provides a financial incentive to keep the forest intact, instead of being converted to environment and atmosphere damaging cattle ranches and palm oil plantations, a leading cause of deforestation.

The beauty of the PEFC certification is that it lets everyone fight climate change by giving them the opportunity to buy products made from sustainable forests, from toilet paper and notebooks to wood flooring and grocery packaging.

PEFC's efforts highlight that we can't just complain about climate change. We need to be proactive. Why do we always have to blame Mother Nature instead of owning what we have created? We build housing in flood plains and in areas that have been prone to wildfires since dinosaurs walked on earth. The crazy part is we don't apply knowledge we already have. We have been waterproofing wooden boats forever but the production housing we build in wetland areas still use house wraps designed for falling rain and not rising water. Our houses invite the embers right into the combustible attic space using vent screen designed to defend rats, not flying embers.

When are we going to start supporting innovation that can defend homes from rising water and homes from wildfires? All these applied science and technologies are now available in today's supply chain. It's our insurance underwriters that need to take the product decisions away from our builders by lowering their risk premiums when they use best practices fire-prone areas.

After the Northridge, California earthquake in 1994, we embraced new seismic engineering that is now the law. So when are we going to encourage resilient technologies against weather related events? Houses do not need to be built on poles to become waterproof. When are we going to make houses more resilient to defend wildfires from consuming thousands of homes each year?

We need to adopt new approaches, encourage and support technological innovations, learn lessons from indigenous knowledge that has been passed down through the ages, and also be prepared for the unexpected.

Chapter Seven:
EMBRACING
GOOD FIRE

———

Wildfire is a natural and necessary part of many forest, woodland, and grassland ecosystems. Fire suppression has caused an unnatural buildup of small trees and coarse woody debris that serve as fuel. At the same time, climate change is increasing the heat that drives wildfires. The National Park Service uses prescribed burning and managed wildland fire to reduce fuel loads. These actions also serve to reduce the risks of fire under continued climate change.

~ NPS.gov

———

W hen COVID-19 upended California's spending plans, the budget for wildfire prevention was gutted. A $100 million pilot project to outfit older homes with fire-resistant materials was dropped. Another $165 million earmarked for community protection and wildland fuel-reduction was slashed to less than $10 million.

Then August of 2020 ignited.

According to the California Department of Forestry and Fire Protection (CAL FIRE)—an emergency response and resource protection department—in 2020 wildfires scorched more than four million acres—a modern record—killed thirty-one people, and damaged or destroyed more than ten thousand structures. It was a record-shattering wildfire year and cost California an estimated $1.3 billion in emergency firefighting costs. The California Department of Forestry and Fire Protection's total spending was more than $3 billion. Such numbers make penny-wise, pound-foolish sound fiscally responsible.

The MSN news service reported: "The numbers highlight the

enormous chasm between what state and federal agencies spend on firefighting and what they spend on reducing California's wildfire hazard, a persistent gap that critics say ensures a self-perpetuating cycle of destruction."

Michael Wara, director of the climate and energy policy program at Stanford University's Woods Institute for the Environment believes the system in place for managing the wildfire problem doesn't work anymore.

"I think the reality is," Wara says, "just like we need to pay for fire protection, we probably need to be paying something like that much money for fire risk reduction in the state."

We also need to rethink our options because the intensification of wildfires has been driven not just by the weather but by forestry practices. In a 2007 research paper, UC Berkeley scientists estimated that prior to 1800, about 4.5 million acres of California burned every year in fires ignited by lightning and Native Americans. The government's twentieth century fire suppression policies put an end to that, producing a long-term fire deficit and fuel buildup across much of the state that Californians are now paying the price for. How that came to be is a lesson in unintended consequences.

History of Fire Suppression Policies

A series of forest fires in the late 1800s caused concern that forest fires threatened future commercial timber supplies and watersheds. In 1891 those concerned enabled conservationists to convince the federal government to start setting aside forest. The US Forest Service was established in 1905 to manage what we now called national forests. And an integral part of that management was fire protection.

Then five years later several forest fires burned three million acres in Montana, Idaho, and Washington in just two days—known then as the Big Blowup—which had a great effect on national fire policy. Local and national Forest Service administrators were adamant that the destruction could have been prevented if they'd had adequate resources of men and equipment on hand. They also convinced Washington politicians and the public that only total fire suppression could prevent such an event from occurring again, and that the Forest Service was the only entity capable of carrying out that mission, which had two goals: preventing fires and suppressing a fire as quickly as possible once one started.

On the prevention side, by seeing fire itself as the enemy, the Forest Service as official policy opposed the practice of light controlled burning, even though generations of ranchers, farmers, and timbermen had used the practice because it improved land conditions and the quality of soil. So while those who worked the land understood how to care for it, politicians and those in charge of managing the forests had limited understanding of fire's important ecological role in forest health. Forest Service administrators took the stance that all fire in the woods was bad because it destroyed standing timber. To help push that message, in 1944, the Forest Service introduced Smokey the Bear to help "educate" the public.

On the suppression front, the Forest Service built networks of roads, communications systems, lookout towers, and ranger stations. To protect both federal and non-federal lands, Congress passed the Weeks Act of 1911 at the urging of the Forest Service. The law established a framework between the federal government and the states for cooperative firefighting. By offering states financial incentives to fight fires, the Forest Service directed the national fire policy.

Bad fire seasons in the early 1930s during the Great Depression made fire suppression an even greater urgency. In 1933 the federal government created the Civilian Conservation Corps (CCC) to build fire breaks and fight fires. Two years after that the Forest Service established the *10:00 a.m. policy*, which mandated that every fire should be suppressed by 10 a.m. the day following its initial report, regardless how remote or how contained. Other federal land management agencies quickly joined the campaign to eliminate any fire from the forest landscape. The efforts to eliminate any burning was supported by new technologies and resources such as airplanes, smokejumpers, and fire-retardant chemicals.

Suppressing fires as quickly as possible remained the official policy until the 1960s when scientific research contradicted the underlying theory as incorrect, demonstrating that in fact fire played a positive, essential role in forest ecology. So in the early 1970s the Forest Service did an environmental about-face and instituted a new policy to let fires burn when and where appropriate. It began with allowing naturally caused fires, such as from lightning strikes, to burn in designated wilderness areas. Although that approach was widely second guessed after the 1988 Yellowstone fires, time proved science right.

The fire burned 1.4 million acres in the greater Yellowstone ecosystem, including 793,880 acres of the national park. It started thanks to a perfect storm of conditions and events: High winds, drought, lots and lots of fuel, and low humidity. The Yellowstone fire was actually multiple fires that at the time prompted the largest firefighting effort in American history to date. It was extinguished thanks to a snowstorm the second week of September.

Images of the charred national park sent shockwaves through the American public as well as people worldwide, so it was inevitable that the Park Service was roundly criticized for mostly letting nature take its course. But ecologists were not particularly upset. Earlier in 1988, fire and plant experts had reported that the Yellowstone area through millennium had experienced massive fires every 200–400 years. The last major burn had occurred in the 1730s, so the area was due for another major burn.

And within a few years, visitors were able to see the park coming back to life, with new, healthier growth and more fertile soil. Interestingly, the local wildlife from birds to grazers to predators, had mostly survived the fire intact and were now benefiting from the new growth. The entire ecosystem was rejuvenated, attracting more visitors than ever who were fascinated to see nature's cycle of sustainability up close and personal.

"The '88 fire was a reaffirmation that fire was what this landscape needed, and it was a great opportunity to get that message out," says John Cataldo, Yellowstone's fire management officer.

In the 1970s the National Park Service adopted a policy in Yosemite to minimize suppression and limit prescribed burns. Forty years later in 2016, two researchers from the University of California at Berkeley published an assessment of the national park's health. They found the *let nature (mostly) take its course* approach had made Yosemite more resistant to catastrophic fire, with more diverse vegetation and increased water storage, in meadows created in areas cleared by fires.

One of the study's authors said, "When fire is not suppressed, you get all these benefits: increased stream flow, increased downstream water availability, increased soil moisture, which improves habitat for the plants within the watershed. And it increases the drought resistance of the remaining trees and also increases the fire resilience because you have created these natural firebreaks."

To this day both Yosemite and Yellowstone maintain the strategy to let nature play out unless human life is in danger or if any cultural or historical landmarks are threatened. The parks also suppress any fires caused by humans, whether accidentally or on purpose.

"The plants and animal communities have been adapting and evolving for millennia to different compositions of vegetation," Cataldo says, "so as long as we keep Yellowstone wild, we'll be doing the right thing."

Writing in the *Monterey County Weekly*, David Schmalz's account of his experience in the aftermath of a Northern California fire echoes the Yellowstone and Yosemite stories.

"This past weekend, on a camping trip with friends in Big Sur, I hiked through part of the burn scar of the 2020 Dolan Fire on the Kirk Creek Trail, and what I found was a vibrant landscape of thriving chaparral and blackened redwoods that were scarred, but for the most part, very much alive. It was a reminder that, irrespective of how wildfires start, fire is natural. It was also a reminder that when we let fires burn, the impact of the next fire in that area won't be as devastating."

According to the Forest Service's report on the Dolan Fire, which started in August of 2020 but wasn't fully contained until December, states: "Recent burn scars, such as the 2016 Soberanes [fire], have helped reduce fire spread. In the 2008 fire scar areas, where shrub growth is recent, the fuels are not receptive."

Schmalz added, "One would think that by now the Forest Service has learned its lesson: fire suppression is futile, and only makes future fires potentially more damaging to the landscape. But a recent article in *The New Yorker* made clear that's not the case."

For example, it revealed that Randy Moses, chief of the Forest Service who has called the United States' wildfires a national crisis, also promotes a policy of full suppression and scaling back prescribed burning. It just shows that even though the 10:00 a.m. rule was officially rescinded decades ago, it still informs policy. The *New Yorker* article also reported:

> The US Forest Service and the Department of the Interior employ some fifteen thousand wildland firefighters, who are directed to prioritize fire suppression; as a result, ninety-eight percent of all wildfires in America are extinguished before they become large. But preventing fuels from burning today preserves them to burn tomorrow. As the stockpile grows, fires burn longer and with greater ferocity. In California alone, an estimated twenty million acres—an area the size of Maryland, Massachusetts, and New Jersey

combined—would need to burn to eliminate the so-called fire deficit created by a century of suppression. Federal agencies acknowledge the problem, but bureaucratic risk aversion and budget constraints, among other things, have stalled the adoption of new approaches, leaving America both burning and fire-starved.

Rather than continuing down the path of the 10 a.m. policy, many forest experts are more vocally advocating for *good fire*, a term used to describe moderate fires that not only take away the fuels that create megafires but they are also critical to the life cycle of several tree species. For example, ponderosa pines are considered fire dependent. And according to the National Park Service, "communities burned naturally on a cycle of one every five to twenty-five years. This frequent fire burned the grasses, shrubs, and small trees, and maintained an open stand of larger ponderosa pine trees. Fire is essential to shaping and maintaining ponderosa pine forests."

The National Forest Foundation lists other trees whose life cycles count on fire. "Typically, species that regenerate by re-sprouting after they've burned have an extensive root system. Dormant buds are protected underground, and nutrients stored in the root system allow quick sprouting after the fire. Shortleaf pines (also occasionally called southern yellow pines) employ this technique."

The actual seeds of many plants in fire-prone environments need fire, directly or indirectly, to germinate. These plants produce seeds with a tough coating that can lay dormant, awaiting a fire, for several years. Whether it is the intense heat of the fire, exposure to chemicals from smoke or exposure to nutrients in the ground after fire, these seeds depend on fire to break their dormancy.

The National Forest Foundation goes on to note: "A combination of factors has come to limit and alter historic fire regimes. This has had a cascading effect on which species are present in certain ecosystems. Without the right kind of fire regimes, some trees simply can't reproduce, and overall forest health can be negatively affected. At the same time, unnaturally severe fires can destroy forests, even those that have adapted to fire. Fortunately, land managers are realizing the value of reintroducing controlled fire in ecosystems where it existed historically, embracing it as a tool rather than fighting it as a threat. That's a welcome change for the forests, trees and other plants that depend on fire to thrive."

Like they say, everything old is new again. Long before Europeans arrived in North America, indigenous tribes managed their forests using good fire. And it wasn't unique to the land we now call the United States. Indigenous people around the world used good fire to keep forests healthy. And many in the forest industry believe it's the only way to make forests truly fire-resilient. It's a tool that can restore our environment, our ecosystem, and maintain the strength and health of our planet. Experts estimate that forty million people live in fire-prone landscape that has been largely deprived of rejuvenating flames for more than a century.

In modern times, firefighters reduce this fuel through planned burning or through cutting out overgrown vegetation. But it hasn't kept up. Overgrown and dead vegetation has crowded beneath trees across tens of millions of acres in California. The state has increased its efforts to thin more acres, but it will take decades at the current pace to treat the vast area that will be needed to buy down the fire danger that has piled up.

The Politics of Minimizing Climate Change

Whether you are a believer in climate change or not, it's clear all humans need to support—now—new proven science that defends the future of our environment. We need healthy forests and to prevent megafires by embracing the value of natural fire and rejecting 10:00 a.m. and other suppression policies. Forest management problems have filled our land with fuel. Climate change has helped to dry it out and create the perfect storm for megafires.

Nobody is suggesting states cut firefighting budgets; what experts are suggesting is to invest more in prevention and mitigation. But there is a sticking point that so far has proven difficult to overcome. Letting fires do what they do in remote places like Yellowstone or Yosemite, which are isolated wildlands, is a no-brainer. The problem is every year development encroaches on wildlands, bringing with it the variables of homeowners and commercial landholders, creating what is called the wildland urban interface (WUI).

The WUI is defined as "the zone of transition between unoccupied land and human development." It is where structures and other human development meet or intermingle with undeveloped wildland or vegetative fuels. WUI fires are becoming ever more commonplace, and not just in the western states. Fire experts have determined more than forty-six million residences in 70,000 communities in the United States are at risk for WUI fires. Having people's lives at risk complicated the equation, but many of the suggested solutions have been questionable.

For example, around 2020 lobbyists for the timber industry and local leaders convinced Congress to fund the creation of three hundred thousand acres of buffer zones around some communities in Northern California, including Quincy.

According to an NBC report, "But what was supposed to take five years ended up taking fifteen years. Costs mushroomed amid environmental lawsuits over logging on forest lands. Backers of the Quincy program hoped that the logging would offset the costs of cutting and hauling everything away. That cost more money, but it spared the area from the air pollution from controlled burns. In the end, the final assessment concluded that although the program limited fire risk, the project suffered too many setbacks to be a model."

Not everyone has such a benign response to clear-cutting three hundred thousand acres of forest. In 1996 Chad Hanson co-founded the John Muir Project. He developed his interest in national forest protection after hiking the complete 2,700-mile length of the Pacific Crest Trail from Mexico to Canada with his older brother in 1989.

"It's really just a misguided prescription," Hanson said of the buffer zone, calling it a transparent effort to justify destructive logging on federal land.

He stresses that *mechanical thinning*—the term for clear-cutting—won't be enough to stop megafires, which are climate and weather fueled. Hanson is adamant that best way to live with wildfires it to let lightning-caused fires burn.

"We need more fire, small ones and big ones," Hanson said. "What we need to ask ourselves is how do we keep fires from burning homes? How do we protect and save lives in vulnerable communities from wildland fire?"

I can answer that. We can't just focus on one factor or the other. We need to manage for healthy forests and that will help us with our climate goals that are absolutely necessary while also balancing it with public safety. So, how do we protect and save lives in vulnerable communities from wildland fire? Through Mighty Fire Breakers' Locked-N-Loaded Wildfire Defense system.

Chapter Eight:
AN EFFECTIVE LOCKED-AND-LOADED
WILDFIRE DEFENSE SYSTEM

W hen I started working as a technical engineering representative in the lumber industry, I wanted to do more that point out problems. I wanted to create solutions. I knew we had to do more to prevent damage to the environment, lost property, and civilian and firefighter deaths. The best way to do that was to become a technologist and develop systems and products to eliminate fire's ability to advance on wood structures. On average, wildfires have produced 8–15 percent of CO_2 releases in the United States. Also, mold contamination can occur following a building fire, which presents an additional health danger.

So in 2008 I pivoted into fire science after learning about a scientist in Malaysia who had invented some inhibitors. With his approval I started to blend and improve on his initial formula and then took them to American laboratories to get accreditation. That end product is what became the basis for a new way to fight wildfires on many fronts through my new company, M-Fire Suppression.

Our Mighty Fire Breaker applied fire science uses a clean, safe fire inhibitor—not a fire retardant. And we have proved over and over that our technology can put out wildfires faster and safer, protecting the environment, homeowners, and our brave fire fighters. This same fire inhibitor can put our tire fires, diesel fire, and lithium battery fires. For years it has been the proven way to put out deep underground peat fires in Indonesia.

Spraying our inhibitor on dry vegetation *before* a fire eliminates the ability of a wildfire to use the brush as fuel, which in turn protects homes and buildings, as well as nearby wildlife. Treated areas enable fire fighters to put fires out faster. And any areas that have burned can be misted once and never reignite, so we also limit toxins from being washed into our streams, lakes, and underground water. There is no odor and Mighty Fire Breaker's fire inhibitor is safe around landscaping, pets, and humans.

Our system is technologically state-of-the-art but also extremely user-friendly, using water as the delivery agent. It's a sprinkler system—some units are placed on the roof of the house or building and some on nearby slopes—that is connected to a tank containing our inhibitor. If a wildfire should happen and the land is threatened or residents are ordered to evacuate, all the property owner has to do is turn the system on, and the house and all the growth around it will get saturated with the product, which will provide protection for a month or two.

I created this system because I was tired of seeing so many people lose their homes. Now you have something to defend your property when a fire is calling on you. The inhibitor compound clings to the vegetation and prevents the fire from advancing. And what also sets this system apart is that it supports root growth, and there is no clean up required like is needed for the red fire retardants currently used. The chemistry that I'm using is not a phosphate, like what they drop out of the planes. I'm using tripotassium citrate, one of the few fully biodegradable and non-toxic flame retardants that support new growth.

M-Fire Holdings has started installing the proactive home defense systems in Southern California, but I have to admit it's been a more of a challenge to break through the red tape and status quo than I expected. So I've taken my system and done demonstrations at fire training academies in front of hundreds of fire officials, who are the biggest skeptics because they think they have seen it all. But when I show them what happens when using my system during a wildfire burn or a wood-frame building burn, their jaws all drop and they become believers. Unfortunately, they then all say the same thing: *Good luck getting upstairs.*

I did a demonstration early on at a controlled burn in Ventura County, going out the day before to treat the hillside. I laid down giant letters cut out of plywood that said *Save Cali* and sprayed the dry vegetation sticking out of the plywood boards. Then we showed up the next day at eight o'clock in the morning, and all these young firefighters arrive with handheld hoes. All day long the fire chief has them digging fire breaks, checking their endurance.

At 4:30 p.m. when the typical California late afternoon wind started blowing, the chief set the hill on fire. The minute the fire reached the dry vegetation and the plywood letters that were sprayed the day before, the fire went out, so the sprayed letters *Save Cali* appeared as everything else around it burned and turned black.

The fire chief couldn't believe what he had just witnessed, so he put more fuel on the areas that didn't burn and relit them. Our inhibitor put it right out.

I looked at all the young firefighters and asked them, "What would you rather fight a wildfire with a hoe or a backpack full of chemistry?"

Their answer was unanimous. They all yelled: *"The backpack!"*

About a year and a half after that demonstration, a major fire broke out in Ventura County. I watched it on television, feeling so frustrated, angry, and helpless. The news reports were saying the fire was heading toward the 101 freeway, and everyone was worried: *Oh my God, what's going to happen if it jumps the 101?* Now, they threw every resource at that fire—planes, helicopters, retardant, fire fighters with hoes—and still it jumped the freeway and took out more than five hundred houses.

We need to forget the planes. They're a waste of resources and a waste of finances because by the time they drop a blob of fire retardant, go fill up again, and come back, the fire has advanced. And helicopters hold such a relatively small amount of water, it amounts to a video-op for media.

Here's the reality. For all the wildfire fighters' training, the terrain of many wildfires makes it difficult for crews to reach the front lines. But if we had four-thousand-gallon tanker trucks able to spray our inhibitor three hundred feet, we could design firebreaks where we know the fire is moving toward. More than that, we could spray a firebreak in fire season *before* a fire even happens. As long as it didn't rain, our inhibitor would still be active a month later, clinging to vegetation without killing the vegetation. What would anybody have to lose to just try it?

My efforts to promote this product have taken me from local Chambers of Commerce, county council members, and all the way to the federal government. I also find myself having to do a lot of education, explaining that not all wildfires are bad. They are necessary. For example, the forests that have died from beetle infestation actually need to burn. We need to control fire through an inhibitor such as ours so we can use fire to help better manage our forests.

But we also need to better control fires once they start. It's crazy expecting firefighters to beat back a fire driven by Santa Ana winds with a hose. All we have to do is look at the photos and videos of the recent destruction up and down the West Coast. The gels don't work, the foams don't work. And they don't give you enough time to get them installed. The only time we call for the plane to swoop in is as a last resort when we've already been beaten. Despite their cost—often running into the millions of dollars each—they have relatively little impact. Plus the phosphate coming down in blobs the way it does has an environmental impact, despite the FDA continually saying it is environmentally safe.

In December 2018, *Organic Lifestyle Magazine* reported this about Phos-Chek, the red retardant currently used. "Although the formula is kept secret, the fire retardant is composed primarily of fertilizers like ammonium phosphate combined with clay or guar thickeners designed to keep the solution from dispersing in the air. Phos-Chek use in the state of California has multiplied rapidly over the last few years, going from nine million gallons sprayed in 2014 to nineteen million gallons used in 2016. That trend promises to continue, as more than a million gallons of the chemical were used on the Mendocino Complex Fire this year."

It was also reported that while phosphorus is an essential nutrient for plant growth, there can be too much of a good thing. "Excess phosphorus, which remains in the soil for three to five years, causes plants to develop yellowing leaves due to an inability to properly absorb nutrients like iron, manganese, and zinc. It also harms root fungi, interfering with a plant's ability to absorb water."

Unloading nineteen million pounds—more than 8,600 tons—of this phosphate-based fertilizer on our forests and wildlands, is harming native plants.

Andy Stahl, executive director of Forest Service Employees for Environmental Ethics, explains that "phosphate fertilizers like Phos-Chek, can have adverse effects on plants adapted to nutritionally poor soil by increasing competition from invasive species better suited to growing in the newly-fertilized soil. For this reason, the US Forest Service bars aerial fire retardant from being used in critical habitat of many threatened or endangered plants."

And according to AccuWeather, it can "kill fish and make a waterway toxic with ammonia and phosphate if dropped over or near an aquatic area."

On so many levels a system like M-Fire's is only common sense. Think about a place like Malibu, California. The officials there shouldn't want to be the definition of insanity by doing the same thing and expecting a different outcome. They shouldn't want to have to continually rebuild Malibu. Instead they should be open to new technologies that defend their city, their homes from the next fire that will happen at some point. If they did, Malibu could be a model city for the rest of the world.

The First Line of Defense: Builders

While wildfires get all the attention, fire defense needs to start at the beginning. We have to create a balanced build going forward, so it can't be said enough: we need to support everyone involved to promote new, better, more sustainable, resilient ways to reduce loss while we reverse the enormous damage done to this planet since the Industrial Revolution. We need to reimagine how we construct homes and buildings; we must use more renewable and sustainable materials and systems.

It is accepted internationally that greenhouse gas abatement in construction is achieved by increasing timber usage and could be attributed to National and State carbon accounts. Quantification and recording of embodied carbon and sequestered carbon at a building level offers a useful tool to compare structures and designs and optimize the use of materials. Storing the sequestered carbon in a building would recognize the benefits of sustainable building products to sequestered carbon, and its potential to deliver a carbon storage effect.

Globally initiatives are being implemented to reduce CO_2 releases into the atmosphere. In the complete forestry cultivation process, the use of lumber for housing, industrial, and consumer products—and the balanced build considerations of CO_2-intensive production of steel and concrete—produce wood products which sequester CO_2 for the lifetime of those products.

Wildfire defense is a key consideration in assuring sequestered CO_2 within inventories of timber-framed construction stays sequestered. The level of carbon stored at a building-scale is about 50 percent higher for timber frame than masonry, and significantly higher for CLT, as much as 400 percent the concrete structure. The embodied carbon of a typical residential timber-framed house reduced by 1.7–3.2 metric tons carbon dioxide equivalents (CO_2e) compared with a functionally-equivalent masonry house. In addition, the timber-framed houses also stored 2.0-4.2 times more CO_2e sequestered carbon in the structural elements than a masonry house.

In comparing concrete framed and cross-laminated timber apartment blocks, a greater differential was seen, with a 12.8 to 18.0 metric tons CO_2e reduction per flat in embodied carbon of the structure and an increase of 12.4 to 17.3 metric tons CO_2e stored sequestered carbon. It is estimated that every 100,000 timber-framed and CLT system homes built would result in (net) additional storage of sequestered carbon of around one metric ton of CO_2e. The California Cap and Trade 2021 vintage price sits around $18/tonne. That's a potential $18 million per year valuation for every 100,000 homes not lost to wildfire in California.

Regardless of whether individuals, communities, or government agencies believe the earth is warming, it is undisputed that wildfires and unplanned fires in structures release CO_2 into the atmosphere and create enormous hazards to all involved. I, and every stakeholder of Mighty Fire Breaker, are working toward advancements in fire protection and defense locally and globally to reduce the effects of wildfires and released CO_2. We need to support those architects, builders, and even insurance providers that promote new, better, more sustainable, resilient ways to reduce loss.

It just seems that the people who are most impacted and the entities that are most impacted would be willing to at least try a product like Might Fire Breaker. It's hard to believe homeowners wouldn't be interested in seeing a demonstration for a product that could significantly reduce—or even eliminate—the homeowner's vulnerability to fire, and in turn result in reduced premiums. But it remains a battle getting people to pay attention and understand—or maybe even believe—there is applied science today that can save their homes, loved ones, and communities from fires.

Instead, look at the way we defend ourselves from wildfires: We hire thousands of young guys and give them a handheld hoe to go and dig a firebreak, often too close to the fire itself, as was seen in the 2020 Northern California fires when several firefighters lost their lives.

The Mighty Fire Breaker Process

While nobody has yet figured out how to eliminate wildfires, Mighty Fire Breaker can stop them much more quickly than current efforts. Our process is thorough and scientific.

First our team flies a drone over the home we are going to defend and identify where the biggest risk would come from. That is the direction we will direct the system's sprinklers. But risk isn't just from the main flames. Before we install the Locked-N-Loaded sprinkler system, the property owner must be willing to remove other risks and take proactive measures. These include:

- Trimming dry vegetation and trees that support a wildfire advance.
- Installing gutter defender so dry, dead leaves are not an easy ignition.
- Retro fitting all roof, soffit, gable, and crawl space vents with Vulcan vents, which block flying embers.
- Clearing any debris, junk, or old wood that's easily from around the house and any outer buildings.
- Clearing all dead or dry vegetation from around propane tanks and replacing with noncombustible gravel or concrete pads.
- Removing and relocating anything stored under wooden decks.
- Either removing Eucalyptus trees or taking off limbs at least ten feet off the ground.

● If the window screens are not metal already, replacing them

When all those preventive measures are checked off the list, we install the system. Then we will spray trimmed vegetation around the structure twice a season when we come to inspect the roof top system. And wood shake roofs and wooden siding must be sprayed twice a season between June and September. We also spray dry decks and fences to remove a wildfire's ability to advance and to prevent hot embers from igniting anything next to or on your structures.

After the personalized defense plan and quote is given to the property owner, they can decide to do all the work or have us help them with the labor. Once everything on the report is complete we will install the GPS-tracked Locked-N-Loaded Defense System.

Our No Shortcut program is designed to save your property and reduce risk for the insurance provider. We come out twice a year to check on the system and do whatever preventive spraying is needed. While there we take pictures of the property and all the fire prevention measures and provide them to your insurance provider to show the home and property is truly defended.

Because it uses a gas-powered pump, our system is self-contained. Once primed, it does not count on the homeowner's water or power. After the system saturates structures and areas around structures, the fire inhibitor remains effective for months, even when dry. Our inhibitor breaks the free radical chain in fire, creating a fire break that makes it extremely difficult for embers and fire to remain ignited.

Our program and system do not count on private firefighters who use foam and handheld tools to save homes. We count on proven applied fire-accredited science. In that way, we hope to help property owners who have lost their insurance get reinstated because our program lowers the insurance providers' risk of loss. Now whether or not the insurance companies do the right thing will likely depend on state legislatures.

Insurance for Insurers

What's happening in California is a microcosm of what every is affecting American living in an area that could even remotely be affected by a wildfire. In coast states, if you bought a home on the beach, you knew going in that your homeowners' insurance policy could be complicated and expensive—or in some cases almost impossible to get. But today many homeowners who live in areas now considered brushfire zones never thought they would be at risk for insurance cancellations.

Only hurricanes and floods outpace wildfires in terms of the largest payouts for claims filed on homeowners policies. So not surprisingly wildfires are one of the fastest growing sectors of high-risk homeowners insurance. And not just in California but Arizona, Colorado, Idaho, Montana, Nevada, New Mexico, Oklahoma, Oregon, Texas, Utah, Washington, and Wyoming. Insurance companies are increasingly declaring parts of states and even entire counties as high risk.

In California, homeowners across the state have taken measures to help protect their homes from wildfires and satisfy risk-averse insurers, but many are seeing their homeowners' policy premiums jacked up anyway or even cancelled.

In 2020 Insurance Commissioner Ricardo Lara backed a bill that would have required insurers to renew policies for homeowners that met state standards for defending their home against wildfire. It died in committee after strong opposition (lobbying) from insurers, and Lara has since said he plans to use regulatory powers to create an insurance program that incentivizes California homeowners to take mitigation measures.

Both consumer groups and insurance companies seem to agree that they support creating an insurance-based mitigation program. But they unquestionably disagree on how such a system would work.

The Santa Cruz County newspaper, *Good Times*, wrote: "The fundamental questions that need to be answered are: Who will certify that homeowners meet mitigation standards? How will those standards be determined? Will insurers be allowed more flexibility to adjust rates in conjunction with such a program?

"As Lara has pointed out, when car owners show their insurance company that they're a safe driver who avoids accidents, they're usually given a discounted rate. California homeowners who demonstrate that they've made their home more immune to wildfire are hoping for the same deal on their insurance policy."

Everyone at M-Fire agrees with Commissioner Lara, and that is why we have invested in accredited testing to show our program truly hardens homes. It completely outperforms what private insurance providers' private firefighters use to spray. It will also outperform the red fire retardants sold in the big box stores or any gel because our chemistry eliminates ember ignition and the fire's ability to advance, even after it dries. As of this writing, we were still hoping to get a meeting with Insurance Commissioner Lara to prove we have the documented wildfire defense he has been looking for.

It's relatively easy to make a rebuilt home previously lost to wildfire very fire-resilient. It's the thousands of homes that have not been consumed yet that need real, no short cut, affordable applied fire science to support them against the next wildfire attack. I have been yelling *Fire!* for the last fifteen years or more, and I know our technology will have a positive impact on nature's future.

To date Mighty Fire Breaker has protected twenty-million-square-feet of residential and business buildings. In agriculture, our patent-pending CitroTech natural fire inhibitors have protected thousands of acres of vineyards and farms from devastating wildfires.

We cannot hem and haw about what to do when whole communities are burning down, releasing alarming amounts of greenhouse gas because we can't put out wildfires faster. We have to learn about new applied science and technologies to spray wildfire fuel under tree canopies in our forest instead of using antiquated methods of back burning that only produces more greenhouse gas. The powerful firefighters' unions need to put down the ancient tools used to dig fire breaks and embrace new chemistry that gets the job done faster and safer and more effectively.

In addition to the Locked-N-Loaded system, we have also developed treated retention wood mulch. So instead of blending the mulch with water, it's blended with our fire inhibitor. The treated mulch can last six months as a fire break on the side of roads used for wildfire exiting. All those country roads where people die in their cars trying to escape and evacuate could be widened with a fire-treated retention mulch that would last throughout the hot, dry months of peak wildfire season.

The backend of that is then defending all the carbon sequestered in the trees so it's not released back into the environment during out-of-control wildfires, which brings us full circle back to M-Fire Holdings' prevention system.

The Politics of Wildfires and CO_2 Sequestration

I'm a big supporter in all clean energy efforts but until we address fire and smoke they will have no gain. Big lumber and clean energy are still all about profit and dividends for shareholders. In addition to clean chemistry that shuts down wildfire and greenhouse gas production, we need to support carbon trading and reward lumber producers and builders to motivates the change. They say: *Put a price on carbon.* Well, nature is screaming that it can't pay that price anymore.

By September 2020, ten of the largest fires in modern history were all burning at once in California. That year wildfires destroyed more than four thousand buildings, forced hundreds of thousands of people to flee their homes, and scorched more than 3.2 million acres across the state.

For however much the state is doing, it isn't anything close to what is necessary. Thinning and prescribed burns generally cover only a tiny fraction of what the Little Hoover Commission recommended.[2] In 2018,

[2] https://lhc.ca.gov/

the state passed a law dedicating $1 billion over five years to wildfire prevention. In 2019 Governor Gavin Newsom signed a package of fire bills that included another $1 billion for preparedness and emergency response.

That's still not at the levels needed.

According to a report in *Technology Review*, the Little Hoover Commission recommended "cleaning out 1.1 million acres a year. That would still take two decades, and require a lot of workers and money. Prescribed burns on forest and park lands can cost more than $200 per acre, while thinning can easily top $1,000, depending on the terrain. So the total costs could range from hundreds of millions of dollars to well above a billion per year.

"Still, that's a fraction of the costs incurred by out-of-control wildfires. To take just one example, the devastating Wine Country Fires in October 2017 did more than $9 billion worth of damage in a single month. Battling wildfires on US Forest Service land runs more $800 an acre."

In August 2020 California reached an agreement with the US Forest Service to treat a million acres per year for the next two decades. However, it's just a fingers-crossed hoped-for goal; it's not a binding law, and there's no guarantee of more funding.

Michael Wara, director of the Climate and Energy Policy Program at the Stanford Woods Institute for the Environment says, "The funding is key. As is a clear line of accountability if they don't actually follow through."

Politicians rarely agree on public policy, but there are issues that Americans in general and California is particular need to work together on, such as healthy forests, protecting our most vulnerable communities, and providing sufficient and dedicated wildfire funding.

After the Northridge Quake the federal government funded the retro fitting of all freeways overpasses and the building industry had to embrace new lateral strength engineering. We need to face the fact that construction design implemented in the post-World War II housing boom has also contributed to wildfire's destructiveness. Gable vents, roof vents, soffit vents and crawl space are major contributors to loss of homes during wind-driven wildfire events. Not to mention that we now build homes around areas that have regularly burned since dinosaurs roamed the earth. To spend all the funding to thin forests and vegetation is as ridiculous as how we spend most of the funding to drop fire retardants from expensive planes when there are roads to spray from.

Thinking it's preferable to thin dry vegetation instead of funding a tax credit to harden all the production houses that now are built with zero wildfire resilience. Real hardening is not about just trimming. We need to retrofit all rat screen vents. We need to use technology like Locked-N-Loaded that will not the fire from advancing instead of dropping retardant from a plane once the wildfire is already out of control.

There are solutions. There are always solutions. It's a matter of finding them and getting them to the people who can actually implement them.

Steve at another arson fire set
during construction.

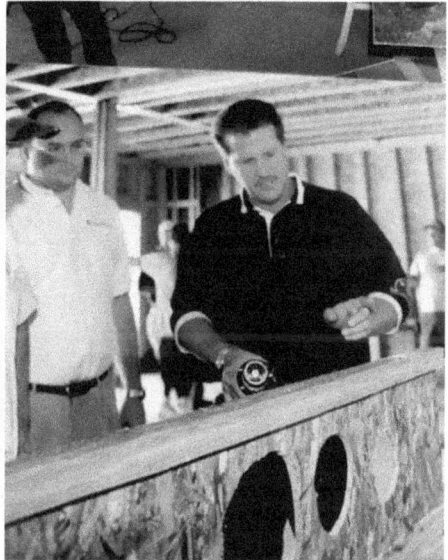

Steve was an award-winning technical
engineering rep for Trus Joist, the first
company to bring engineered wood
products to the national supply chain.

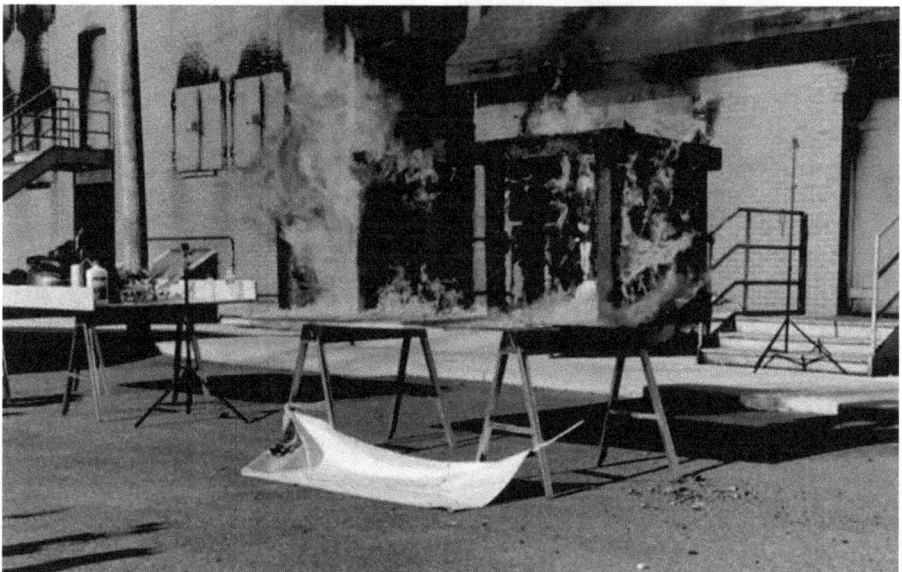

Wildfire and building fire demos in New York City

Nature Is Screaming; Are We Ready to Listen?

Teaching wall panel plants how to fire defend prefab walls.

Steve in Denver inspecting a construction site arson fire.

Steve trains and licenses general contractors all around the Western US for the No Shortcut Wildfire Defense Program and Locked-N-Loaded systems.

PREVENTING FIRES AT CONSTRUCTION SITES

Live interview about arson attacks in Oakland

Chapter Nine

GREEN TECHNOLOGIES THAT HAVE
HEARD NATURE SCREAMING

———

In recent years, as more consumers have become increasingly concerned about climate change and the environment, a lot of companies have jumped on the green bandwagon. The only problem is, just saying you're green doesn't mean you really are, a practice called greenwashing. A CNBC.com report found that a study of more than one thousand products made green claims that were either "demonstrably false or that risk misleading intended audiences."

Well, in the case of Mighty Fire Breaker, we truly are and can back up our claims that we develop, produce, and distribute innovative fire inhibiting and extinguishing products that do not contain harmful ingredients and polluting technologies. Our core product, Mighty Fire Breaker MFB-31 with our patent-pending CitroTech plant-based chemistry, is an eco-friendly, proactive wildfire defense solution for property owners and is also designed for mass timber producers and prefab factory-built housing developers.

We can provide safe and effective fire inhibition without harming people, pets, wildlife, or our planet because we start with renewable carbohydrate raw materials derived mainly from corn, which are transformed through fermentation and other biochemical processes into readily biodegradable and ecologically safe ingredients that are used to formulate our fire inhibitor products. We use no phosphates, no ammonium, no fluorosurfactants—zero toxic chemicals, so our products will not contribute to nutrient pollution, which the United States EPA characterizes as "one of America's most widespread, costly and challenging environmental problems, with far-reaching impacts on human health." So when you use our fire inhibitor products in your home, workplace, and within your community, you know that you are joining us to create a healthier and more sustainable world for generations to come.

A Grassroots Effort

As with most things, change comes from the bottom up. You can't really rely on federal politicians who are often beholden to lobbyists and big businesses to lead the charge. People need to write their local and state representatives and demand that contractors rebuilding houses destroyed by fire be mandated to fire-treat any surviving framing and all new lumber used in construction before they wrap and set windows and dry the buildings.

Houses sprayed with MFB-31M and that have a Locked-N-Loaded Wildfire Defense System have a radically reduced risk of property loss or personal injury. And it gives property owners peace of mind if they should ever have to evacuate.

Lumber is still the best material to build with because it's natural, eco-friendly, inexpensive, and sequesters carbon like trees in our forests do while also supporting a healthier ozone layer, which in turn makes for a healthier planet. Most importantly when lumber is fire protected, it can last for hundreds of years while continuing to support a greener environment.

"Wildfires cause damage to human lives and billions of dollars in property loss each year, yet there has been shockingly little innovation in this space," said Peter Diamandis, founder of XPRIZE, a non-profit that designs and hosts public competitions to encourage new technologies that benefit humanity.

Park Williams, a climatologist at UCLA and a leading expert on how climate change has intensified the problem of wildfire, agrees. "Fires are outrunning us. We're trying harder than ever to put them out, and they're continuing to win, more and more, every year. And it really isn't for lack of effort. Even when we know it's been stupid policy to fight every single fire, we're still trying as hard as we can to do that. The root problem that got us into this mess is thinking that because humans put a man on the moon and created the Internet, we should also be able to dominate wildfire. And in reality we can't. It's much larger than us when it really gets going."

This is why government-funded community engineered wildfire defense systems would fight climate change better, save taxpayers hundreds of millions of dollars, make things so much safer for firefighters, and defend the future of our environment better than any program to date. We need the government to treat this wildfire crisis like they do a pandemic because our firefighters are doing the best they can, but it's nowhere near enough until we defend communities as our military does with missile defense systems. The applied science is here and ready to prove it can stop wildfires dead in its tracks if Locked-N-Loaded systems are there and deployed before wildfires arrive.

To accurately convey the scale of wildfire risks and impacts at the community level, we need to focus on the numbers that tell the story. The number of structures destroyed, rather than the number of acres burned, is a more telling measure of the broad social, economic, and community impacts from wildfires. And since 2005 more than 100,000 structures have been destroyed by wildfires, resulting in an untold number of fatalities, evacuations, and personal losses. Wildfires are burning thousands of structures each year, and that trend will likely continue to increase until we look at better proactive systems beyond just trimming.

The Need for Green Wildfire Defense Engineering

We need to stop saying wildfires are a part of nature when we are talking about losing entire communities. If it was just timberland, it would be one thing and hardly make the news. But many of the community losses do not occur in forests. Fires are spread by embers on the wind, and dry native vegetation fuels in difficult terrains make them hard to stop no matter how many super tanker planes we deploy. Today's wildfires are destroying homes, wildlife, timberlands, and vineyards, as well as polluting our air and waterways.

Communities have used a variety of sources to fund capital projects to pay for operations and maintenance costs as investments in climate adaptation. But only when the federal government listens to nature's screams and funds new green-tech wildfire defense, like Locked-N-Loaded, plants more trees to sequester carbon with a tax credit, and institutes carbon trade to fire-treat the lumber we use will we turn the corner on wildfire carnage and more effectively sequester CO_2 and defend the carbon storage that fire releases back into our environments.

At heart I'm a lumberman. We have figured out how to fire-treat lumber at 600 a minute coming out of a sawmill planer—and it doesn't even change the color. My method is so affordable that a simple carbon trade would offset the value add. That, planting more trees to support housing, and spraying under tree canopies, is the future, and the best way to get governments to support that is with a carbon trade.

Thousands of architects have taken our American Institute of Architects (AIA)/US GreenBuild Council-approved courses to learn about applied fire science that lowers all risks of loss. If the mass timber and lumber industry would fire-treat lumber to defend the carbon stored in its wood fiber so it's never released back into our environment, this would finally quiet nature's screams, reduce risk premiums, and support carbon trading, leading to bigger and greater reforestation.

Bringing Builders Onboard

The US building industry needs to support applied fire science for residential homes that are built in wildland fire regions, just like the big national builders are doing on five-story wood-framed, high-density housing around the country. Communities are not only resisting unchecked building, but residents also want to know that developments that are built are part of the wildfire and environmental solution.

In February 2021 the city of Pittsburg, California, approved a 1,650-house development on more than six hundred acres of WUI land. California environmentalists opposed to development in semi-urbanized areas filed a lawsuit.

The Save Mount Diablo website explained. "Aside from all the negative impacts to open space, wildlife, views, and the new Thurgood Marshall Regional Park, we said that Seeno had not submitted anywhere near the amount of information necessary to determine what the project actually was. There was no precise description, no specific map of development, no infrastructure details; Seeno just approved the concept of 1,650 houses on top of Pittsburg's hills on a big patch of map. We said that this violates the California Environmental Quality Act and that there was not enough information to allow decision-makers to fully understand what was being proposed. Additionally, the project would fail to gain the other approvals necessary because Seeno did not provide enough information."

Attorneys for the park district argued that the development "poses a severe, hazardous fire risk. The East Bay Hills are especially vulnerable to catastrophic fire because of hot and dry fall seasons and steep, wind-conducive topography."

The Local Agency Formation Commission (LAFCO) for Contra Costa County agreed and rejected Seeno's application to advance the project. The reality is that since 2015 wildfires have wiped out tens of thousands of homes in Wildland Urban Interface zones, the semi-rural hillsides just outside cities such as Santa Rosa, Chico, Santa Cruz, and Redding. These WUI areas are home to 11.2 million Californians, and the widespread destruction and displacement from recent blazes is prompting questions about whether builders should continue adding housing in them. And residents are pushing back like never before, and fire dangers are one of the major talking points.

If builders want to develop even a small portion of WUI land, they need to show they have embraced modern green tech that is effective and environmentally safe, especially in an area like Mount Diablo because, as the park attorneys noted, hot, dry winds that blow from the northeast every spring and fall, make a housing vulnerable to potentially devastating wildfires.

"The Diablo winds can fan the flames of small sparks into wildfires that have been observed to move down from a ridgetop in thirty minutes, expand to one square mile in an hour, and consume hundreds of residences in one day."

Matt Regan, who heads up public policy for the Bay Area Council, adds: "We have a small developable footprint in the Bay Area compared to the rest of the country. We have done a very good job of preventing sprawl and a bad job of making it easy to build in areas where building is appropriate. It's really hard to do infill development in the city, but the places that are easy to build are likely to be more prone to fire."

The first builder we worked with was in New Jersey years ago after they lost some high-density projects to arson. After ninety days of due diligence, they signed a ten-year contract with us to fire-defend 100 percent of their five-story, high-density projects. Many other national high-density builders are now using that program because it helps lower their risk and their insurance premiums against arson activity. Why not wildfires?

Wildland arsons account for 20 percent of brush fires. Most arsons happen during the daytime and are set for many reasons, including vandalism, crime concealment, extremism, profit, excitement, or revenge. A wildland arsonist typically looks for opportunity such as areas with dead and dying brush, trash, or abandoned furniture. But nobody needs to be fearful of wildfires—whatever their cause—when they become proactive with new science and understand the nature of wildfires.

Other Emerging Wildfire Technologies

In addition to our on-the-ground fire defense, other technologies are emerging in the fight to protect property, people, and the environment from the consequences of wildfires, which are not just a West Coast phenomenon. In North America alone the area that fires affect has doubled between the late 1980s and 2021. It only makes sense to utilize advancing digital technology in the effort to control wildfires. In 2013 a team at the University of California Berkeley started a project called the Fire Urgency Estimator in Geosynchronous Orbit (FUEGO) that uses drones outfitted with special thermal imaging or infrared cameras to monitor and track a wildfire's progression. If it becomes a significant threat, the system dispatches air tankers and ground firefighters to the fire's location to hopefully control it before it spreads.

Drones are valuable because they can be flown into areas that manned aircraft cannot access, especially at night. In British Columbia, drones are now commonly used for mapping and hotspot detection, providing overnight maps that can be used by fire and evacuation crews upon first light.

Advancements in robotics have resulted in a firefighting robot. Developed by Howe and Howe Technologies, Thermite is a robot platform that can also be fitted with a robotic arm or cameras for reconnaissance or configured as a bulldozer.

"Thermite robotic firefighters are one of the most capable, durable firefighting robots on the market. Designed to mitigate life-threatening situations, these tools provide fire suppression, situational awareness, and intelligence gathering to first responders. Operated by remote belly-pack controllers, users are provided a real-time video feed allowing them to traverse hazardous terrain and push obstacles from their path while withstanding extreme elements. Thermite robotic firefighters are critical to have on your side in high-risk, dangerous environments."

By sending in a combination of robots, firefighters can not only put out fires but also assess the situation and search for survivors. Wildfires can burn so hot that firefighters are unable to get close enough to extinguish them. Now robots can help fight wildfires even in the most extreme conditions.

Smart early-warning sensors can be used to detect and measure the level of CO_2 and check for unseasonably high temperatures, indicating the possible presence of fires in the area. What makes them "smart" is that they are connected to and can exchange data with other devices and systems over the Internet or other communications networks—collectively known as the Inter of Things (IoT). Another advantage is that these connected devices require minimal power.

But all these are just tools. What really has to change is our mindset. In July 2021 Timothy Ingalsbee, a former federal firefighter who now heads Firefighters United for Safety, Ethics, and Ecology, told CBS News that firefighters need to adopt a new approach when confronting the most dangerous wind-driven wildfires that leapfrog containment lines by showering flaming embers a mile or more ahead of the main inferno. For one thing, it's better to build more fire-resistant homes and devote scarce resources to protecting threatened communities while letting the fires burn around them.

"We have these amazing tools that allow us to map fire spread in real time and model it better than weather predictions. Using that technology, we can start being more strategic and working with fire to keep people safe, keep homes safe, while letting fire do the work it needs to do, which is recycle all the dead stuff into soil."

It's not just homeowners and businesses in the United States that are adopting a new proactive approach to wildfires; it's happening all over the world. For example, in early 2022 we shipped our first international order to Sierra Leone, Africa, where our MFB-31-CitroTech product will be used to protect the Beach Café, a newly built eatery and bar located at the most popular beach in the heart of Freetown, Sierra Leone's capital. Our product will be used to protect the wooden structure of the Café and its traditional thatch roof, which otherwise would be extremely flammable and impracticable.

The Beach Café, the country's first eco-friendly restaurant, is the creation of the Propel Organization, a Sierra Leone non-governmental organization dedicated to cleaning the beaches and protecting the environment. All profits from the café support environmental initiatives, such as establishing a sea turtle sanctuary near the café where the turtles can nest and lay eggs. Propel selected MFB-31-CitroTech not just because of its ability to resist fire ignition but because research shows it is the only fire suppression chemistry retardant that is truly green and safe with no impact on aquatic life.

Sierra Leone is battling a rash of fires that have recently destroyed numerous buildings in densely populated urban areas like Freetown. Because power is so uneven, there are many interruptions of service. Then when power is restored, voltage spikes can cause devastating electrical fires. Hospitals are ill-equipped to handle burn victims, and fire trucks rarely arrive in time. Week-in and week-out, lives are lost, and homes are destroyed. Propel is planning to showcase the Beach Café as the first fire-hardened building in Sierra Leone and in West Africa. We are especially proud that MFB-31-CitroTech is the first-ever green fire suppressant chemistry used on the African Continent.

According to Belinda Botha, CEO and co-founder of Propel, "West Africa is experiencing an unprecedented building boom. Although

Sierra Leone is a developing country with enormous challenges, the demand for environmentally friendly, safe products like MFB-31 is significant. Everyone is concerned about fire, and builders, government agencies, and international organizations all want to be on the green side of the solution."

We invite homeowners; local, state, and federal government officials; firefighters and their unions; insurers; and builders to adopt a new mindset and embrace a truly green, proactive fire defense technology that can benefit us all by lowering the risk of property loss and promoting a healthier, sustainable world for generations to come.

IN CONCLUSION

Everyone has a role to play to help Mother Nature stop screaming, and we cannot afford to wait.

The World Needs to Get Water Smart

Our applied fire science is ready to prove it can put out fires much faster while also using much less water than traditional firefighting methods. The amount of water consumption currently used in fighting wildfires is alarming—hundreds of millions of gallons. We have to embrace change *now* on how fire departments fight wildfires, consuming so much water when there is a new proven science that can defend and shut them down with less water consumption and less toxic runoff.

If the water shortage, toxic runoff contamination, and the production of greenhouse gas are part of the Climate Change Initiative, then the federal government should challenge Might Fire Breaker's claims now. Shutting down wildfires with less water using environmentally friendly chemistry with no ammonia and no phosphates that also reduces greenhouse gas production should be supported by the big fire departments. This is simply promoting more sustainable applied science that will defend all of us better.

ESG Investing

Environmental, social, and governance (ESG) investing allows you to put your money to work with companies that strive to make the world a better place. ESG investing relies on independent ratings that help you assess a company's behavior and policies when it comes to environmental performance, social impact, and governance issues.

Bank of America estimates that in the next twenty years there will be more than $20 trillion of asset growth in ESG funds, equal to the current size of the S&P 500.

Hank Smith, head of investment strategy at the Haverford Trust Company, says, "At its core, ESG investing is about influencing positive changes in society by being a better investor."

Develop a Sense of Urgency

In February 2022 the United Nations published a report that confirmed what I've been saying for years: uncontrollable wildfires are intensifying with the rise in greenhouse gas emissions, and communities in general and property owners in particular are not prepared for the escalating potential damage.

The report forecasts that by 2030 extreme wildfires will increase by 14 percent, 30 percent by 2050. In seventy-five years the risk will increase by 50 percent.

"Due to a combination of climate change and land use change, there has been a dramatic shift in wildfire patterns worldwide. Some areas, such as the Arctic, will probably experience a significant increase in burning by the year 2100. Tropical forests in Indonesia and the southern Amazon are also expected to experience increased fires if greenhouse gas emissions continue unchecked."

As Peter Moore, an author of the UN report and a forestry fire management consultant, says, "We're spending an enormous amount of money on suppression, particularly in the developed world. That solution has run its course. In my opinion, it's reached its limits."

This apocalyptic peek into the future doesn't have to happen. But we need to change our approach. And that starts with using the applied fire science already available with CitroTech and the Locked-N-Loaded system. We can break the cycle of wildfires releasing CO_2 that causes global warming that in turn is a factor in increasing numbers of wildfires, which releases even more CO_2 by trusting our proven applied fire science.

I cannot scream it often enough or loud enough: there is now new applied fire science that puts fires out faster than ever before and if adopted, it will be a game-changer in the efforts to combat and control wildfires and reducing carbon emissions.

There is no time to waste.

Coming Soon ...

The Big Wood and Little Wood children's illustrated book series. An old-growth tree and a young sapling will teach children about wildfires and the importance of trees and forests to the environment. With science-based content presented in engaging, age-appropriate language, Big Wood, Little Wood will also show how we can all help make Mother Earth more sustainable.

```
*  9  7  8  1  7  3  5  2  7  2  4  7  4  *
```